Recent Advances in Grapes and Wine Production - New Perspectives for Quality Improvement

Edited by António M. Jordão, Renato Botelho and Uroš Miljić

Published in London, United Kingdom

Recent Advances in Grapes and Wine Production - New Perspectives for Quality Improvement
http://dx.doi.org/10.5772/intechopen.100883
Edited by António M. Jordão, Renato Botelho and Uroš Miljić

Contributors
Teng-Zhen Ma, Faisal Eudes Sam, Bo Zhang, Evangelos Beris, Moustafa Selim, Despoina Kechagia, Alexandra Evangelou, Valeriu Cotea, Elena Cristina Scutarasu, Camelia Elena Luchian, Lucia Cintia Colibaba, António M. Jordão, Tatiane Otto, Renato Botelho, Luiz Biasi, Uroš Milijić, Ana C. Correia

First published in London, United Kingdom, 2023 by IntechOpen
IntechOpen is the global imprint of INTECHOPEN LIMITED, registered in England and Wales, registration number: 11086078, 5 Princes Gate Court, London, SW7 2QJ, United Kingdom

British Library Cataloguing-in-Publication Data
A catalogue record for this book is available from the British Library

Additional hard and PDF copies can be obtained from orders@intechopen.com

Recent Advances in Grapes and Wine Production - New Perspectives for Quality Improvement
Edited by António M. Jordão, Renato Botelho and Uroš Miljić
p. cm.
Print ISBN 978-1-80356-323-7
Online ISBN 978-1-80356-324-4
eBook (PDF) ISBN 978-1-80356-325-1

Meet the editors

António M. Jordão is an Associate Professor of Enology at the Polytechnic Institute of Viseu (School of Agriculture), Portugal, where he coordinates the Viticulture and Enology technical course, and is a member of the Chemistry Research Centre. He graduated in agro-industrial engineering from the Agronomy Higher Institute of the Technical University of Lisbon, obtaining a master's degree in food science and technology and a Ph.D. in agro-industrial engineering from the same institution. He is associated with several research projects involving grape and wine phenolic composition, and wine technology. He is the scientific editor of seven books on grape and wine composition and winemaking and has published 50 scientific papers in peer-reviewed journals and 20 book chapters. He is involved in developing consultancy services for wine companies and evaluates research projects for National Innovation Agencies in several countries.

Renato V. Botelho has been an Associate Professor of Viticulture and Fruit Crops at the State University of Midwest of Paraná (UNICENTRO), in Guarapuava, Paraná State, Brazil, since 2003. He graduated in agronomy from the Faculty of Agronomic Sciences – UNESP, in Botucatu, São Paulo State, Brazil, obtaining an MSc and Ph.D. in Agronomy from the same institution. His postdoctoral projects are in the Fruit Crops section of the Agronomic Institute of Campinas, São Paulo State, Brazil and in the Agrarian and Veterinary School of the University of Bologna, Italy. His research interests are viticulture and organic-system fruit crops. He has published 110 scientific papers, is a reviewer for numerous international journals, and advises several research funding agencies.

Uroš Miljić is an Assistant Professor of Wine and Alcoholic Beverages Technology at the University of Novi Sad, Faculty of Technology, Novi Sad, Serbia. He graduated in food biotechnology and obtained his Ph.D. in technological engineering from the same institution. He is the author/co-author of 31 articles in peer-reviewed scientific journals, three book chapters, and 29 papers in scientific conferences. He is based in an accredited wine and spirits analysis laboratory, and his research interests are in different fields of wine technology, grape and wine composition, and spirits technology. He is a certificated sensory evaluator of wine quality and a member of expert groups in the Serbian Ministry of Agriculture, Forestry, and Water Management set up to establish the geographical indication of Serbian wines.

Contents

Preface **IX**

Chapter 1 **1**
Introductory Chapter: New Challenges and Innovations in Grape and Wine Production
by António M. Jordão

Chapter 2 **9**
Adaptability of Different International Grape Varieties in Diverse Terroirs: Impact on Grape and Wine Composition
by Tatiane Otto, Renato Botelho, Luiz Biasi, Uroš Miljić, Ana C. Correia and António M. Jordão

Chapter 3 **37**
Overview of the Esca Complex as an Increasing Threat in Vineyards Worldwide: Climate Change, Control Approaches and Impact on Grape and Wine Quality
by Evangelos Beris, Moustafa Selim, Despoina Kechagia and Alexandra Evangelou

Chapter 4 **61**
Enzymes and Biochemical Catalysis in Enology: Classification, Properties, and Use in Wine Production
by Elena Cristina Scutarașu, Camelia Elena Luchian, Lucia Cintia Colibaba and Valeriu Cotea

Chapter 5 **93**
Low-Alcohol and Nonalcoholic Wines: Production Methods, Compositional Changes, and Aroma Improvement
by Teng-Zhen Ma, Faisal Eudes Sam and Bo Zhang

Preface

The grape and wine sector has probably inspired more research and publications than any other area of food and agriculture. Through their passion for vine and wine, a great many scientists have not only contributed to the development of viticulture practices but have also enabled advances in winemaking technology. In addition, the impact of climatic changes on grape production in recent years has stimulated important lines of research analyzing how vines adapt to and resist the new environmental conditions affecting several wine regions. Each applied development in grapes and wines has led to improved control of the physico-chemical and sensory qualities of the different wine products.

During the life cycle of the vine, as well as during all stages of grape development, different factors affect the development and composition of the grapes: climate, soil, wine-growing practices, and the genetic potential that each grape variety presents. Numerous technological innovations in the production process also contribute decisively to the wine quality. The general goal of this book, therefore, is to concisely summarize viticulture and enology innovations and their impact on grape and wine quality.

This book comprises five chapters. The introductory chapter summarizes the new challenges and principal innovations in the production of grapes and wines in recent years. Chapter 2 focuses on the latest science concerning the adaptation of grape varieties in different environments and its impact on grape and wine characteristics. Data on international grape varieties introduced in several wine regions and the native grape varieties are compared. Chapter 3 discusses esca and its impact on vine productivity and grape quality in vineyards worldwide; this increasing threat to global viticulture is causing significant losses in terms of reduced yields, declining or wilting vines and shorter productive life for vineyards. The authors also suggest the possible role of climate change in the spread of the disease. Chapter 4 examines the significance of enzymes in winemaking and their contribution to the development of new strategies for optimizing wine production, highlighting the link between biochemical processes involving enzymes and the quality of wine as a final food product. Finally, the authors of Chapter 5 consider low-alcohol and non-alcoholic wine production, focusing on the principal techniques used in wine dealcoholization, their impact on the wine's phenolic and volatile composition, and on wine sensory characteristics.

The authors of these chapters are international researchers currently involved in research and innovation in different dimensions of grape and wine production. This book is not only for technicians actively engaged in the field, but also for students and other professionals interested in recent innovations and discoveries in the fascinating world of viticulture and enology research.

It was with great pleasure that we accepted the opportunity offered by IntechOpen to assemble and edit this book. We are greatly indebted to the authors who have generously shared their scientific knowledge and experience through their contributions to this book.

António M. Jordão
School of Agriculture,
Polytechnic Institute of Viseu,
Viseu, Portugal

Renato Botelho
State University of Midwest of Paraná,
Guarapuava, Paraná State, Brazil

Uroš Miljić
Faculty of Technology,
University of Novi Sad,
Novi Sad, Serbia

Chapter 1

Introductory Chapter: New Challenges and Innovations in Grape and Wine Production

António M. Jordão

1. Introduction

According to the Report of Global Market Trajectory & Analytics [1], in 2022, the global market value for all wine styles (still, sparkling, and fortified) is expected to reach US$326.6 billion and keep growing at annual rates above 4%. Thus, grape and wine production is a very relevant agricultural activity and represents an important economic sector in the international trade. For wine, after the year of 2020 where there was a large trade disruption world over, the lifting of restrictions due to the COVID-19 pandemic has set the world wine export market on its path to reconciliation. According to the OIV statistics published in 2022, it was possible to obtain in 2021 a value of 111.6 mhl for world wine exports. This was the largest exported volume of wine ever recorded in history. At the same time, world wine exports in 2021 have increased by 4% compared with 2020 and have boosted even more in terms of value with 34.3 billion EUR, registering a yearly increase of 16% [2]. This large economic increase results from a global area of vineyards estimated at 7.3 mha in 2021. This area compared with 2020 shows a very slow decrease of −0.3%. In addition, all this world area under vines refers to the total surface area planted with vines for all purposes (wine and juices, table grapes, and dried grapes), including young vines that are not yet productive.

It is important to note that grape and wine sectors are not only relevant in economic terms but also historically and culturally. In fact, since antiquity, grape is one of the earliest domesticated fruit crops and has been widely cultivated and appreciated for its fruit and wine. Moreover, there is evidence that grapevine cultivation and winemaking dating back to at least 5800 BC [3]. According to Myles et al. [4], grapevine cultivation emerged in the Near East before spreading to Europe and subsequently for other parts of world. In recent years, this historical and cultural richness has also been the object of a strong economic use, namely through wine tourism.

Therefore, despite the economic relevance and long historical past, the production of grapes and wine faces a high number of challenges that lead to an increasing commitment to innovation and sustainability of the entire production chain. Thus, there are several challenges, such as a change in production methods through a reduction in the use of crop protection products and an increase in biocontrol solutions and biostimulants in the different practices of viticulture. On the other hand, the production of wines using a proper identification of the coadjutants used and the reduction in the use of several products with a negative impact on human health, namely the use of sulfur, are other challenges. However, one of the most important challenges of wine industry is related with climate change, particularly through global warming and its impacts on grapevines and their characteristics. In fact, global warming is responsible for many of the problems facing winegrowers around the world. The impact of climate change is currently responsible for several problems in all world regions, such as early grape harvests that are becoming increasingly common. In this case, grape harvesters must work in extremely hot weather conditions, resulting in more frequent breaks and production losses and, at the same time, higher sugar content in the grapes. This last consequence induces a production of wines with higher alcohol levels and at the same relevant changes in the wines' aromas. In addition, early bud break (as early as March) is weakening the vine stock, exposing it to a greater risk of frost until April or even May. Also, frequent droughts and reduced water availability have led to an increase in vine plant destruction. Furthermore, there are also new challenges related with human resource for winegrowers. Thus, in many regions, particularly in Europe, there is a reduction in the available labor force. Consequently, the hiring of foreign workers is an increasing trend, especially for the works in viticulture and during the grape harvest. The future at this level could involve the increasing use of vine robots, thus reducing the necessary work force, but also allowing for an increasing precision viticulture. The use of vineyard robots, which is still uncommon practice around the world, offers several advantages. In that case, robots make tasks less difficult, optimize working time (winegrowers can devote their time to activities with more added value), and increase the profitability.

Finally, new challenges related with the sustainability and circular economy are presented to the wine industry. If compared with other chains, the wine industry is considered of low environmental impact [5]. According to OIV (International Organization of Vine and Wine), sustainable vitiviniculture is defined as a "*global strategy on the scale of the grape production and processing systems, incorporating at the same time the economic sustainability of structures and territories, producing quality products, considering requirements of precision in sustainable viticulture, risks to the environment, products safety and consumer health and valuing of heritage, historical, cultural, ecological and landscape aspects.*" Thus, from this definition, the sustainability of wine sector includes diverse aspects of organic, biodynamic, and integrated production, but at the same time also incorporates the history, the culture, the landscape, and all intangible aspects that characterize the wine production and consumption. In addition, the progress in grape and wine production also has a strong component in the circular economy, where the use of waste and its recovery also present increasing challenges. In a simple way, the circular economy is based on a general idea where waste coming from an activity should be passed to another activity with an important valorization on an infinite loop. Thus, from the vineyard and from the wine production process, various residues can be valued, for example, pruning residues, stalks, pomace, and lees. From these residues, several advances have emerged through the introduction of various technologies, making it possible to develop and obtain various products, such as pellets, biomass, alcohol, biogas, grape seed oil, tartaric acid, and several bioactive compounds (resveratrol, tannins, etc.) used in food and pharmaceutical industries.

2. Innovations in grape and wine production

Without innovative developments over innumerable generations, wines as we know them would not exist. Over the last few years, several innovations have been introduced in the production of grapes and wines. In viticulture, the innovations produced have included the introduction of new production techniques (including different viticultural practices and a better use of water resources) with a view to improving the profitability of grape production and dealing with climate change, the use of new strategies to combat diseases and pests, the development of hybrid grape varieties (including new varieties) well adapted to the new environmental conditions, and also the introduction of new technologies linked to precision viticulture. All these developments try to answer to four fundamental objectives: improvement in the quality of the grapes, reduction in the production costs through mechanization, protection of the environment, and response to climate change.

In terms of soil management, there are several works where a combination of different techniques, such as chemical weeding, soil tillage, and cover-cropping [6, 7], has been developed. In addition, it is well known that an adequate nitrogen supply of the grapevines was proved to play a key role in plant fertilization, and at the same time, nitrogen deficiency could impair the wine quality [8]. In addition, high grape nitrogen deficiencies also affect fermentation kinetics and wine flavors [9]. Thus, several innovative works ranging from plant biology to factors linked to N regulation have been conducted to contribute to the implementation of sustainable practices in the vineyards.

Currently, the management of available water is essential for the sustainability of agricultural activity and consequently for the viticulture. Thus, several innovations have been introduced in terms of rational water management in vineyards. These innovations involve the use of drought-tolerant and drought-resistant rootstocks, a correct canopy management, an adequate irrigation strategy, the use of different sensors for better water management (e.g., the use of electromagnetic induction sensors), and the introduction of remote sensor technologies [10]. In addition, several studies reported that moderate water restriction in vineyards is also favorable for the wine quality [11].

For vine training, a special interest was given to the winter pruning, keeping in mind the respect for the sap flows and trying to limit the expansion of several diseases, namely the wood diseases. Thresholds of leaf/fruit ratios were established and the canopy management during the summer, such as leaf removal and shoot tipping, was adapted accordingly. Related with vine diseases, in recent years there has been a rapid development of strategies to combat these sanitary problems in vineyards based on the use of biocontrol products. Thus, the practical combat has included the use of natural products of mineral, plant, or microbial origin, the implementation of strategies related with the antagonistic microorganisms, and the use of plant defense inducers. All these strategies have contributed to reduce viticulture's dependence upon synthetic fungicides [12, 13].

Also related with the use of new plant material more resistant to disease and drought, the use of hybrids in viticulture is currently being discussed. The introduction of this new grapevine plant material introduces new perspectives to reduce the use of pesticides and increase the adaptation of vines to climate change [14].

In the winery, the recent developments of sensor technologies offer the possibility to control all the relevant parameters for a correct winemaking process and to guarantee a high quality of wine produced. These new technologies help winemakers to have a set of information in real time, not only during fermentation but also during wine aging, which help them to make the best decisions throughout the wine production process. Recently, new knowledge about the fermentation process has led to the introduction of strong innovations in wine production processes. In the past, *Saccharomyces* spp. yeasts were almost the only option for used during alcoholic fermentation. This was due to the high ability to metabolize all grape juice sugar into ethanol. However, several results proved that also the use of *non-Saccharomyces* strains can improve the wine quality. In this context, the use of these new strains of yeasts contributes positively to improve the wine acidity, aromatic complexity, contributing at the same time for low levels of acetic acid and ethanol produced, among other positive effects [15]. Thus, in the past years, the main manufactures started to commercialize dry non-*Saccharomyces* strains in the market of oenological products containing different yeast species (e.g., *Torulaspora delbrueckii*, *Schizosaccharomyces pombe*, and *Pichia kluyveri*).

In recent years, due to greater control over alcohol consumption in some countries (related with health problems associated with the consumption of alcoholic beverages), as well as due to new trends in wine consumption, several techniques have been developed for the reduction of ethanol content in wines with excessive alcohol content [16]. In addition, climate change has also contributed in some warmer regions to the production of grapes with excessive amounts of potential alcohol. Also related with the effects on health because of the consumption of alcoholic beverages, several studies have been carried out for the production of wines with reduced sulfur content (or even sulfur free), but at the same time maintaining the wine quality. Thus, several alternative technologies have been developed and compared, such as pulsed electric fields (PEF), high-pressure processing (HPP), power ultrasound (US), ultraviolet irradiation (UV), high-pressure homogenization (HPH), filtration, and low electric current (LEC). All these technologies have been explored with the aim to obtain adequate microbial inactivation and at the same time maintaining the wine quality [17].

To induce better performance to clarify and stabilize wines, new natural, non-allergenic, and non-animal fining agents have been developed by the different manufactures helping winemakers to obtain wines with high quality [18]. At the same time, recent developments in filtration technologies have been introduced to help winemakers to reduce the problems of precipitation of unstable proteins present in white wines after bottling. This problem can cause cloudiness, which is generally considered commercially unacceptable [19].

Finally, for wine aging, different technologies and winemaking practices have also been developed. Several options have been made available, such as wine aging using wood fragments, combination of micro-oxygenation with wood fragments, different options in wine aging on lees, the improvement of wine aging in bottles, and mechanisms in acceleration of wine aging using different technologies (physical methods involving ultrasonic waves, gamma rays, electric fields, nanogold photocatalysis, and high-pressure treatments) [20].

In conclusion, all these challenges contribute to the continuous innovation in the production of grapes and wines and the consequent concern with their quality and at the same time increase in the sustainability in all production chains. Together, consumer and market demands are increasing, thus inducing the necessary investment in constant innovation.

Author details

António M. Jordão[1,2]

1 Polytechnic Institute of Viseu, Agrarian Higher School, Viseu, Portugal

2 Chemistry Research Centre of Vila Real (CQ-VR), Vila Real, Portugal

*Address all correspondence to: antoniojordao@esav.ipv.pt

References

[1] Research and Markets Wine - Global Market Trajectory & Analytics [Internet]. 2022. Available from: https://www.researchandmarkets.com/reports/338680/. [Accessed: November 12, 2022]

[2] State of the World Vine and Wine Sector 2021 [Internet]. 2022. Available from: https://www.oiv.int. [Accessed: September 25, 2022]

[3] McGovern P, Jalabadze M, Batiuk S, Callahan MP, Smith KE, Hall GR, et al. Early Neolithic wine of Georgia in the South Causcasus. Proceedings of the National Academy of Sciences of USA. 2017;**114**:E10309-E10318. DOI: 10.1073/pnas.1714728114

[4] Myles S, Boyko AR, Owens CL, Brown PJ, Grassi F, Aradhya MK, et al. Genetic structure and domestication history of the grape. Proceedings of the National Academy of Sciences of USA. 2011;**108**:3530-3535. DOI: 10.1073/pnas.1009363108

[5] Merli R, Preziosi M, Acampora A. Sustainability experiences in the wine sector: Toward the development of an international indicators system. Journal of Cleaner Production. 2018;**172**:3791-3805. DOI: 10.1016/j.jclepro.2017.06.129

[6] Delpuech X, Metay A. Adapting cover crop soil coverage to soil depth to limit competition for water in a Mediterranean vineyard. European Journal of Agronomy. 2018;**97**:60-69. DOI: 10.1016/j.eja.2018.04.013

[7] Crézé CM, Horwath WR. Cover cropping: A malleable solution for sustainable agriculture ? Meta-analysis of ecosystem service frameworks in perennial systems. Agronomy. 2021;**11**:862. DOI: 10.3390/agronomy11050862

[8] Verdenal T, Dienes-Nagy Á, Spangenberg JE, Zufferey V, Spring J-L, Viret O, et al. Understanding and managing nitrogen nutrition in grapevine: A review. OENO One. 2021;**55**:1-43. DOI: 10.20870/oeno-one.2021.55.1.3866

[9] Loulakakis KA, Morot-Gaudry JF, Velanis CN, Skopelitis DS, Moschou PN, Hirel B, et al. Advancements in nitrogen metabolism in grapevine. In: Roubelakis-Angelakis KA, editor. Grapevine Molecular Physiology & Biotechnology. Dordrecht: Springer Netherlands Publisher; 2009. pp. 161-205

[10] Mirás-Avalos JM, Araujo ES. Optimization of vineyard water management: Challenges, strategies, and perspectives. Water. 2021;**13**:746. DOI: 10.3390/w13060746

[11] Simonneau T, Lebon E, Coupel-Ledru A, Marguerit E, Rossdeutsch L, Ollat N. Adapting plant material to face water stress in vineyards: Which physiological targets for an optimal control of plant water status? OENO One. 2017;**51**:167-179. DOI: 10.20870/oeno-one.2017.51.2.1870

[12] Wurms KV, Chee A, Elmer PAG, Agnew RH, Wood PN. Developing new biologically-based products for control of *botrytis* bunch rot. Part 1: Developing a new natural product for mid-season botrytis control - NP2 moves closer to the market. Wine and Viticulture Journal. 2011;**26**:64-72

[13] Rantsiou K, Giacosa S, Pugliese M, Englezos V, Ferrocino I, Río Segade S, et al. Impact of chemical and alternative

fungicides applied to grapevine cv Nebbiolo on microbial ecology and chemical-physical grape characteristics at harvest. Frontiers in Plant Science. 2020;**11**:700. DOI: 10.3389/fpls.2020.00700

[14] De la Fuente LM. Use of hybrids in viticulture. A challenge for the OIV. OENO One. 2018;**52**:231-234. DOI: 10.20870/oeno-one.2018.52.3.2312

[15] Maicas S. Advances in wine fermentation. Fermentation. 2021;**7**:187. DOI: 10.3390/fermentation7030187

[16] Bucher T, Deroover K, Stockley C. Production, and marketing of low-alcohol wine. In: Morata A, Loira I, editors. Advances in Grape and Wine Biotechnology. London, UK: IntechOpen; 2019. DOI: 10.5772/intechopen.87025

[17] Silva FVM, Van Wyk S. Emerging non-thermal technologies as alternative to SO_2 for the production of wine. Food. 2021;**10**:2175. DOI: 10.3390/foods10092175

[18] Marangon M, Vincenzi S, Curioni A. Wine fining with plant proteins. Molecules. 2019;**24**:2186. DOI: 10.3390/molecules24112186

[19] Sui Y, McRae J, Wollan D, Muhlack R, Godden P, Wilkinson K. Use of ultrafiltration and proteolytic enzymes as alternative approaches for protein stabilisation of white wine. Australian Journal of Grape and Wine Research. 2021;**27**:234-245. DOI: 10.1111/ajgw.12475

[20] Tao Y, García JF, Sun D-W. Advances in wine aging technologies for enhancing wine quality and accelerating wine aging process. Critical Reviews in Food Science and Nutrition. 2014;**54**:817-835. DOI: 10.1080/10408398.2011.609949

Chapter 2

Adaptability of Different International Grape Varieties in Diverse Terroirs: Impact on Grape and Wine Composition

Tatiane Otto, Renato Botelho, Luiz Biasi, Uroš Miljić, Ana C. Correia and António M. Jordão

Abstract

In the last two decades, several international grape varieties from different traditional wine countries such as, France, Portugal, Italy, and Spain have been introduced in several world wine regions, increasing their representation in the world. The introduction of grape varieties in emerging regions with diverse environmental conditions from their natural origin introduces challenges on the adaptability of these varieties in new specific "terroirs," not only in terms of their productivity but also related with the grape and wine characteristics. In addition, it is also important to compare their characteristics with native grape varieties grown in the local regions. On the other hand, climate change has also promoted greater mobility of grapes to new regions, increasing the dispersion of various grape varieties in areas where viticulture was, until now, severely limited. Thus, considering the importance of the dispersion of several grape varieties in regions out of the original provenance, the purpose of this chapter is to present a review of the most recently published data about the adaptation of several grape varieties in different environments and the impact on their qualitative characteristics (including in wines produced). The comparative analysis with some of the native varieties existing in these environments, when applicable, will also be analyzed.

Keywords: adaptability, chemical composition, grape varieties, wine quality, sensory profile, terroir

1. Introduction

The *Vitis* genus (80 species identified) is composed of two sub-genera: *Muscadinia* and *Euvitis* [1]. The *Muscadinia* sub-genus comprises three species, including *M. rotundifolia*. This sub-genus grown in the south-east of North America is remarkably resistant to the main cryptogamic diseases to which most *Vitis vinifera* varieties are prone [2, 3]. However, most cultivated grapevines belong to the *Euvitis* sub-genus. These sub-genera fall under three groups: the American (made up of more than

20 species), the East Asia (comprises about 55 species), and the Eurasian group (composed of one single species, *V. vinifera* L.). For *V. vinifera* specie, there are two sub-species: *sylvestris*, which corresponds to the wild form of the vine, and *vinifera*, referring to the cultivated form [4].

The *Vitis* international variety catalog identifies 21,045 names of varieties, which includes 12,250 for *V. vinifera*. However, this last number includes a considerable number of synonyms and homonyms [5]. Nevertheless, according to Lacombe [6], the number of vine varieties for the *V. vinifera* species in the world is estimated at 6000. The high number of varieties is a consequence of the preservation and transport of vine seeds by farmers, which was a common practice in the past. However, also the interspecific hybridization of *Vitis*, which occurred during nineteenth century until the mid of twentieth century, also contributed to the diversity of genetic material. In that case, the phylloxera crisis had an important role in the creation of high diversity of plant material. Lastly, the natural genetic mutations, which are common in grapevines, also contribute for this diversity.

According to OIV data, grape vineyards (which corresponding to the total surface area planted for all proposes—wine and juices, table grapes, and raisins) covered more than 7.3 million hectares worldwide [7]. Lecat et al. [8] reported that in 2015, the estimation net worth of the wine industry was more than 258 billion euros. However, in 2020, the world wine consumption estimated at 234 mhL and had a decrease of 3% compared with 2019. This decrease could be a consequence of the first year of the COVID-19 sanitary crisis, which induces an asymmetrical aggregate consumption behavior in different countries in the world [9]. Nevertheless, for a global point of view in the last decade, wine sector has undergone considerable changes. One of these changes is related with the grubbing-up of vineyards and restructuring activities. Indeed, some traditionally high-production varieties no longer correspond to the tastes of consumers or the market and have seen a significant decline in their surface area. In addition, as vineyard productivity is strongly related to climate, the tendency observed in last decades for significant changes in climatic conditions, namely for atmospheric temperatures, has been shown to affect grape yield and composition and wine sensory profile. Consequently, the use of different grape varieties of local origin and/or from other wine regions could be a strategy to be followed by winegrowers from different geographical locations [10–12].

In the last 20 years, several authors reported an occurrence of varietal concentration in the world vineyards. According to Anderson and Aryal [13], for instance, Cabernet Sauvignon and Merlot grape varieties have more than doubled their vineyard area. As a result of this situation, Cabernet Sauvignon has been widely studied mostly because its worldwide distribution [14–18]. In addition, numerous studies have shown that grape and wine composition obtained from the different vine varieties depends on several factors that change not only according to the intrinsic potential of each grape variety [17, 19] but also according to the climatic factors, such as sunlight exposition, solar radiation, and temperature [12, 20], soil [21], agricultural practices [22], and also the level of grape ripeness [23]. Thus, winegrowers have traditionally selected and maintained the different grape varieties introduced in the different wine regions, especially the cultivars from other countries and terroirs that best match their specific climates and soil conditions. In addition, recently the wine sector has focused on research and experimental activities about the adaptive capacity of the most economically important grape varieties to climate change in different wine regions. In this perspective, the adaptability of several grape varieties to new regions and climates, as well as, in some cases, the comparative assessment with the native grape varieties from these regions, has allowed us to obtain new data [17–19, 24, 25].

Thus, this chapter focuses on the latest scientific knowledge about the adaptation of several grape varieties in different environments and the impact on grape and wine characteristics. It also approaches diverse comparative data between the international grape varieties introduced in several wine regions and the local native varieties.

2. World distribution of main international grape varieties

According to the OIV database, out of the 10,000 vine varieties known in the world, there are 13 that represent more than one-third of the world vine area [26]. In addition, 33 vine varieties represent 50% of the total vine area. Some varieties are mainly cultivated in a restricted number of countries, such as the Kyoho grape, mainly cultivated in Asia (Japan, China, and South Korea). In an opposite situation, there are other varieties that grow in many countries and usually are called as "international" varieties. One of most important demonstrative examples is the Cabernet Sauvignon grape.

Among the main varieties most cultivated around the world, it is possible to find several countries that are specialized in wine production, such as France, Spain, Italy, Australia, and Argentina, while others are more focused on table and dried grapes production, such as Turkey, Iran, India, or China. However, in the last years, China has shown a great increase in both productions, either wine or table grapes.

Table 1 shows the distribution of the main grape varieties cultivated in the different world geographical areas. As shown in this table, Kyoho grape occupies the largest area of grape vines in the world, although its geographical distribution is restricted to the Asian continent especially in China, which represents more than 90% of the vines area. This table grape variety has been the most produced grape in Japan since 1994, and in South Korea, it accounts for 14.5% of the country's vineyards [27]. For wine grapes, the Cabernet Sauvignon is the most cultivated variety in the world, being distributed in almost all wine-producing countries. From Bordeaux region (France) and derived from a crossing between Cabernet Franc and Sauvignon Blanc variety, Cabernet Sauvignon is the second most-planted vine variety [28]. Today, its vines are widely distributed across the world, covering 5% of the world's vineyards (representing 341,000 ha). It is grown in a great number of countries, such as China, France, Chile, the United States, Australia, Spain, Italy, and South Africa. This grapevine adapts to a wide range of environments with a board phenotypic plasticity (including differential wine sensory attributes). Then, this variety has a high capacity of acclimatization to the different environments and adaptation to climate change [29–31].

For the remaining varieties, such as the Sultanina, Merlot, Syrah, and Tempranillo, it is clear a great geographical distribution being present in several of the major grape wine producing countries. Merlot and Syrah varieties are both from France, the first from the Bordeaux region and the second from the Rhône Valley region. Today, Merlot is present in 37 countries and covered 266,000 ha or 3% of the total world area under vines, while Syrah covered 190,000 ha, and it was grown in 31 countries. Specifically, Tempranillo wine grape is not widely grown outside of Spain, but it may be present in 17 countries. However, 88% of its cultivated area is in Spain.

Although the distribution of the main varieties has spread by the largest grape producers and is very dependent on international varieties, it is important to point out that some countries have obvious dominant varieties in their vineyards, such as Spain,

Grape variety	Skin color	Production destination	Area (ha)	Country origin	Main geographical distribution
Kyoho	Black	Table	365,000	Japan	Japan, China, and South Korea
Cabernet Sauvignon	Black	Wine	341,000	France	China, France, Chile, United States, Australia, Spain, Italy, and South Africa
Sultanina (syn. Thompson Seedless)	White	Table, drying and wine	273,000	Afghanistan	Turkey, Iran, Iraq, Afghanistan, Pakistan, and Central Asia
Merlot	Black	Wine	266,000	France	France, Chile, United States, Italy, and Australia
Tempranillo	Black	Wine	231,000	Spain	Spain, Portugal, and Argentina
Airén	White	Wine and brandy	218,000	Spain	Spain
Syrah	Black	Wine	190,000	France	France, Australia, Argentina, South Africa, United States and Chile
Red Globe	Black	Table	159,000	Italy	China, United States, Spain, Portugal, Italy, and Chile
Garnacha Tinta/ Grenache Noir	Black	Wine	163,000	Spain	Spain and France
Sauvignon Blanc	White	Wine	123,000	France	France, Spain, Italy, South Africa, United States, and New Zealand
Pinot Noir/ Blauer Burgunder	Black	Wine	112,000	France	France, Germany, Italy, United States, Australia, Argentina, and South Africa
Trebbiano Toscano/Ugni Blanc	White	Wine, brandy	111,000	Italy	France, Italy, Portugal, Argentina, and Australia

Table 1.
Distribution of the main grape varieties cultivated in the different world geographical areas. Data obtained from OIV [26].

which has two main varieties, Airen and Tempranillo, that cover more than 40% of the vines area. In China, 44% of the vines are from the Kyoho grape [26]. In addition, there are a few wine countries, such as Italy, Portugal, and Romania that show a quite a diverse varietal distribution, with main varieties not exceeding 9% of the area under vine. These first two countries show an important number of different varieties, especially native cultivars covering 75% of their area of grapevines.

Figure 1 shows examples for a few wine country producers from different geographical origins about the distribution of the mains varieties according to the data obtained from OIV [26]. By analyzing the data presented in **Figure 1**, it is possible

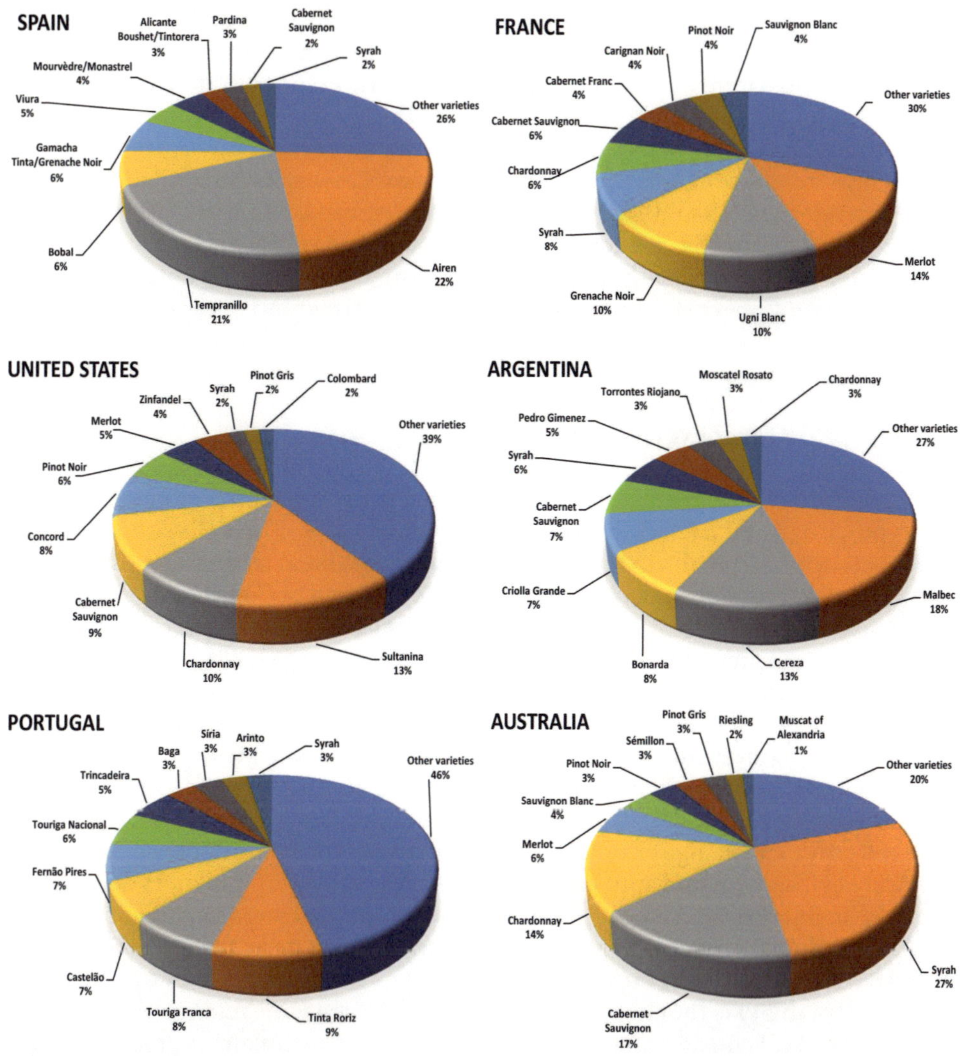

Figure 1.
Distribution of the main varieties in the various wine country producers from different geographical origins. Graphics elaborated by authors using OIV data [26].

to observe that for Spain, one of the biggest wine producers, native varieties are in majority. Airén and Tempranillo occupied around 43% of the vine area. Varieties of foreign origin, namely French, occupy secondary positions, with Garnacha Tinta, Alicante Bouschet, Cabernet Sauvignon, and Syrah being the most representative these varieties.

An opposite tendency is observed for the main varieties cultivated in France. Thus, except for Ugni Blanc, a variety originally from Italy (Tuscany) where it is grown under the name Trebbiano Toscano, all main varieties are from French origin. In that case, Merlot, Grenache Noir, and Syrah are the main varieties, while other varieties very widespread all over the world, such as Chardonnay, Cabernet Sauvignon, and Pinot Noir occupied values between 4 and 6% of the vine area.

An interesting distribution of the main grape varieties is observed in Portugal, a country considered to be one of the countries with the highest varietal distribution, with main varieties not exceeding 9% of the area under vine and at the same time, all main native grape varieties represent around 50% of the vine area. Syrah is the most representative no native variety, and it represents only 3% of vine area. Considering some of the highest wine-producing countries in the "new world," such as Argentina and Australia, the varieties have a foreign origin, namely from France (the majority), but also from Spain, as is the case of the variety Cereza. In Argentina, Malbec and Cereza represent 18 and 13% of vine area, respectively, while for Australia, Syrah (27%), Cabernet Sauvignon (17%), and Chardonnay (14%) are the most representative varieties. Finally, in United Sates, similar tendency is observed, where Chardonnay and Cabernet Sauvignon are the main representative varieties for wine production.

3. Impact of different terroirs on grape composition

According to several authors, it is expected that the combination of plant material (genotype), fungi and bacteria population (microbiome), soil and climate conditions, and all factors related with vineyard management and winemaking affect the quality of grapes and consequently the wines produced. Thus, interactions between all of these factors are usually mentioned as the "terroir" and are finally expressed in the grape composition and consequently in wine characteristics [31–33]. According to Magalhães [34], on the basis of these interactions, the concept of "terroir" has been extensively adopted for the majority of the authors. In fact, all wine regions are characterized by their natural environment conditions, usually related with climate and soil properties, but also depend on human factor. In 2010, the Organization of Vine and Wine issued the resolution VITI 333/2010 with the concept of "terroir" as "an area in which collective knowledge of the interactions between the identifiable physical and biological environment and applied viticultural and oenological practices develops, providing distinctive characteristics for the products originating from this area. Terroir includes specific soil, topography, climate, landscape characteristics and biodiversity features." According to Carbonneau [35], the most important key factor on grape varieties adaptability is the climatic characteristics of each wine region. Instead, Van Leeuwen et al. [36] consider that "terroir" induces all development of grapevine, berry composition and at same time is a key factor that determines the final wine quality, including their typicity and the global characteristics of each wine region.

Thus, it is possible to consider that the adaptability of the different grape varieties, particularly when grown outside their original region is related with a set of factors, namely, atmospheric conditions (temperature, precipitation, humidity, and solar radiation), soil composition and water availability for the plant, potential climate change, pest and diseases, diverse viticultural practices, and varietal/clonal and rootstock selection [37–42]. In the past years, the adaptability of grapevine to different conditions is also related with the development of the modern grapevine breeding, producing numerous hybrid varieties with different characteristics such as cold/hot-resistant, rapid ripening and with resistance to several diseases [43]. These hybrids have made it possible to cultivate vines even in regions where environmental conditions are still a very limiting factor for the development of the vines, as is the case in Nordic countries [44]. Today, also in non-European countries, the use of hybrids is common. For example, in Brazil, several hybrid varieties, such as Moscato Embrapa, Niagara, Villenave, Goethe and Manzoni Bianco and *Vitis labrusca* vines, are usually used with great success for wine production [45]. Also, Cabernet Cortis,

Cabernet Carbon, Bronner, and Regent are other American hybrid vines recently tested in several experimental works in this South American country [46]. However, it is important to note that the preservation of the existent biodiversity related with autochthonous grape varieties is essential to maintain many of the specific and differentiating characteristics of the winegrowing regions. In addition, these varieties could play an important role in the response to climate change and to be used in new opportunities for adaptation to other regions [24, 47].

Cabernet Sauvignon, Malbec, Merlot, Tempranillo, Cabernet Franc, Syrah, and Malbec are several examples of varieties well adapted to different production conditions, namely to very dry and warm climates. Some of them are very cultivated in several countries of southern Europe [48]. In general, these non-native varieties show characteristics associated with own productivity and composition not only related to their own genetics but also develop characteristics that result from their adaptation to the different terroirs where they are implanted.

In Portugal, in the last 20 years, several French grape varieties have been introduced in several wine regions, particularly in the south of the country. For example, Cabernet Sauvignon has been cultivated in different regions, such as Lisbon and Alentejo wine regions with different results and specific characteristics. Thus, Ó-Marques et al. [15] reported no significant differences in grape berry composition, especially related with different general physicochemical characteristics (titratable acidity, pH, estimated alcoholic degree, and berry weight) between Cabernet Sauvignon and Tinta Roriz grape varieties, while other authors found significant differences between several French and Portuguese native grape varieties in other wine regions [49]. In that case, French Cabernet Sauvignon, Merlot, Pinot Noir, and Syrah had higher titratable acidity than two Portuguese grape varieties, Touriga Nacional and Tinta Roriz, in samples collected in a vineyard located in "Douro" region (North of Portugal). In this study also, taking into consideration the phenolic composition, all the French grape varieties assessed had higher total phenols compared with the native grape varieties. For example, Alicante Bouschet (variety from French origin but widely cultivated in the south of Portugal, being officially recognized as a native variety) had the highest values for total phenols and flavonoid compounds (global average value from 0.636 to 0.894 and from 0.584 to 0.834 mg/g of berry, respectively, for total phenols and flavonoid compounds). However, for total anthocyanins, Portuguese native grape varieties (Tinta Roriz and Touriga Nacional) had highest values [49].

Cosme et al. [50] studied the tannin profiles of different *Vitis vinifera* L. red grapes grown in Lisbon region having Portuguese (Touriga Nacional, Trincadeira and Castelão) and French (Syrah and Cabernet Sauvignon) origin. These authors reported that the tannin profile of grape skin was different between cultivars. Thus, the Portuguese Castelão variety showed the lowest mean degree of polymerization (mDP) of proanthocyanidins while Cabernet Sauvignon had the highest mDP. In addition, this French variety had the lowest concentration of total proanthocyanidins in the skins, but this distribution was mainly at the higher mDP (mDP values >30).

It is well known that high temperatures could determine the grape maturation process, accelerating grape berry maturation and at the same time inducing the production of grape berries with higher soluble solids content [51]. In addition, hot temperatures during the day and cold nights induce lower pH values of the grape musts and decrease tartaric and malic acid degradation when compared with regions with hot days and nights. Several authors reported that Syrah, Cabernet Sauvignon, and Alicante Bouschet are grape varieties with minor thermal demands to achieve a normal maturation compared with some native grape varieties, especially taking into consideration the sugars content and titratable acidity in several Portuguese regions characterized by high temperatures [14, 49]. Gordillo et al. [52] reported for warm climatic conditions, which occur in South of Spain (Condado de Huelva D.O.), a high resistance of Syrah variety to these high temperature conditions that occur during maturation.

According to Costa et al. [19], the genetic factor has an important role on phenolic content between the grape varieties and not only the "terroir." According to these authors, the major individual anthocyanin group (monoglucosides) content for Cabernet Sauvignon is independent of the wine region where this variety is produced. Similar tendency was also described by other authors [16, 53], where the biosynthesis of this anthocyanin group is mostly ruled by genetic factors than by climatic conditions. According to Flamini et al. [54] and Sikuten et al. [55], the composition of individual anthocyanins is under genetic control, while agronomic and environmental factors have a greater impact on their total content. In addition, under conditions characterized by dry periods and water deficit, several authors have shown that Cabernet Sauvignon may present lower alcohol level and phenolic content and at the same time higher values of total acidity [56, 57]. Nevertheless, recently other researchers reported under a semiarid climate condition of Israel that Cabernet Sauvignon vines submitted to late irrigation (in last stages of grape maturation) produced grapes with higher color intensities and phenolic content [25]. For different vines from *V. vinifera* and several hybrid cultivars cultivated in the Finger lakes area of New York Stat (United States), Yang et al. [58] reported that Cabernet Franc and Pinot Noir had the highest total phenolic content (values varied from 396.8 to 424.6 mg/100 g) compared with remaining hybrid varieties studied.

In Spain, most vines cultivated are native varieties, occupying more than 50% of the vineyard area. Monagas et al. [59] studied the flavan-3-ol composition of grape seeds and skins from Cabernet Sauvignon and two Spanish grape varieties, Graciano and Tempranillo, cultivated in Navarra region (North of Spain). The higher concentration of flavan-3-ols was observed in Graciano and Cabernet Sauvignon while the lowest values were quantified for Tempranillo variety. This result demonstrates a tendency for an adequate adaptability of the Cabernet Sauvignon under the conditions of this Spanish region, namely in terms of the flavan-3-ols biosynthesis. Several aroma compounds are very important for wines quality, especially for white wines. In this case, terpenes are the source of the fruity and floral aromas found in white wines produced from different grape varieties, such as Riesling, Albariño, Muscat, and Gewürztraminer. García et al. [60] studied the changes in volatile compounds of the most representative white grape varieties cultivated in central La Mancha region of Spain (Macabeo, Airén, and Chardonnay). According to these authors, Chardonnay, a French grape variety that is increasingly being grown in this Spanish wine region, showed higher total acidity compared with the native grape varieties for the same degree of maturity. In addition, several important volatile compounds were quantified in significant values for this French variety. Thus, grape must from Chardonnay had the highest values of *t*-2-hexenel, hexanol, benzaldehyde, phenylacetaldehyde, and benzyl alcohol compared with Macabeo and Airén varieties. These results confirm the good adaptability of this white grape variety to the dry conditions of La Mancha Spanish wine region. Also, in other European regions as in the eastern European countries, several international grape varieties have been introduced. These varieties have been cultivated in parallel with the native grape varieties of these countries. Thus, for example, in Serbia, the most represented native grape varieties are Grašac, Prokupac, Tamjanika, Smederevka, Kadarka, and Bagrina. Apart from this native varieties, international cultivars, such as Merlot, Cabernet Sauvignon, Chardonnay, and Riesling, cover also larger vineyards areas (around 570, 560, 510, and 440 ha, respectively). In the last 10 years, several authors reported comparative data on the chemical composition between international and autochthonous varieties grown in Serbia [61–64].

The vineyards with international varieties have a heterogeneous population and enologists emphasized the importance of clonal selection for getting the clones with best properties for specific terroirs. This is especially important for the varieties with high genetic diversity [65]. The characteristics of a clone mostly depend on the environmental conditions. The production of certified planting material of selected clones is required at the end of process. Vujović et al. [66] studied the evaluation of agrobiological and technological characteristics of three Merlot and Cabernet Franc clones during 5 years (2009–2013) in Grocka wine region (Central Serbia). They observed that individual grape berries from different clones, but also on the same bunch, do not ripe uniformly and the values of the monitored parameters vary significantly. Thus, at the harvest, Merlot (clone 025) and Cabernet Franc (clone 010) had the highest values of total phenols content (7.32 and 10.9 g GAE/Kg frozen weight, respectively) and total anthocyanin content (386 and 1668 mg/Kg malvidin-3-glucoside equivalents frozen weight, respectively). Concerning the individual polyphenols, the following compounds were the most abundant in the both international varieties studied: gallic acid, (+)-catechin, (−)-epicatechin, gallocatechin gallate, (+)-catechin gallate, and rutin [62]. Mitić et al. [61] compared total phenolic content between international grape varieties (Cabernet Sauvignon and Merlot) and two native Serbian varieties (Prokupac and Vranac). The results obtained showed a tendency for the two Serbian native varieties studied and have presented lower values of total phenols (between156 and 158 mg of GAE/100 g of grape) compared with the two international grape varieties (173 and 169 mg of GAE/100 g of grape, respectively for Cabernet Sauvignon and Merlot). Similar trend, i.e., lower amounts of total flavonoids and total anthocyanins, was obtained for Prokupac and Vranac varieties. In fact, Vranac is considered an autochthonous red variety in several Balkans countries, namely from Republic of Macedonia, Serbia, and Montenegro. According to these authors, Vranac grapes also contained similar amount of (+)-catechin as Merlot and (−)-epicatechin as Cabernet Sauvignon. However, the contents of these flavan-3-ols in Prokupac grapes were significantly lower. Opposite tendency was reported by Pantelić et al. [63] where Prokupac variety had the highest content of total phenols in seeds (around 100 mg GAE/g) and skins (around 12 mg GAE/g) compared with several international varieties (Cabernet Sauvignon, Merlot, Cabernet Franc, Syrah, Sangiovese, Pinot Noir, Riesling, Chardonnay, Sauvignon Blanc, and Pinot Gris) cultivated in Central Serbia. These authors also reported for the first time the presence of malvidin 3,5-O-dihexoside in the skins of Merlot, Cabernet Franc, Syrah, Sangiovese, Pinot Noir, and Prokupac grapes, explaining, however, that this compound is not characteristic for *V. vinifera* L. varieties.

Table 2 summarizes several results obtained for general physiochemical and phenolic parameters of several international and native red grape varieties cultivated under different geographical locations.

Also in South America countries, namely in Argentina, Chile, and Brazil, the great majority of vine varieties used are international varieties, having French and Italian origin. In Brazil, most varieties planted are Cabernet Sauvignon, Merlot, Syrah, and Pinot Noir, all of them from French origin, and Barbera, Ancellota, Trebbiano, Riesling Itálico and Moscato Giallo, all of them from Italian origin. Thus, in recent

Parameters	Country/geographical location/reference/grape variety									
	Portugal/Douro region		Spain/Navarra region		Brazil/São Francisco valley		Serbia/Southern Serbia		South Africa/ Western Cape	Australia/ Langhorne Creek region
	Costa et al. [49][1]		Monagas et al. [59][2]		Oliveira et al. [67][3]	Oliveira et al. [68][6]	Mitić et al. [61][9]		Hunter et al. [69][11]	Bindon et al. [70][13]
	Cabernet Sauvignon[a]	Touriga Nacional[b]	Cabernet Sauvignon[a]	Graciano[c]	Touriga Nacional[b]	Syrah[a]	Cabernet Sauvignon[a]	Prokupac[d]	Shiraz (Syrah)[a]	Cabernet Sauvignon[a]
Estimated alcohol degree (%, v/v)	14.0	13.6	—	—	20.87[4]	20.44[4]	—	—	23.93[4]	26.0[4]
Titratable acidity (g/l tartaric acid)	6.7	2.9	—	—	5.22	5.73	—	—	3.93	5.3
pH	3.25	3.75	—	—	3.85	3.78	—	—	4.01	3.48
Total phenols (mg/g berry)	0.474	0.592	—	—	745.37[5]	0.88[7]	1.73	1.56	—	—
Total anthocyanins (mg/g berry)	0.517	0.672	—	—	414.7[5]	0.51[8]	1.20	0.88	601.56[12]	1.88
Total flavan-3-ols (mg/g skins)	—	—	0.38	0.34	—	—	6.32[10]	2.39[10]	—	—
(+)-catechin (mg/g skins)	—	—	0.206	0.104	—	—	4.08[10]	1.17[10]	—	—
(−)-epicatechin (mg/g skins)	—	—	0.171	0.235	—	—	2.03[10]	0.92[10]	—	—

[a]*French variety.*
[b]*Portuguese variety.*
[c]*Spanish variety.*
[d]*Serbian variety.*
[1]*Average values from two vintages (2010 and 2011).*
[2]*Average values from vintage 2000.*
[3]*Average values of four vintages (2014 and 2016) under a tropical semi-arid region.*

[4] *Values expressed as °Brix.*
[5] *Values expressed as mg/Kg.*
[6] *Average values of two vintages (2014 and 2015) under two different altitudes.*
[7] *Values expressed as g/Kg.*
[8] *Values expressed as mg/l.*
[9] *Average values from vintage 2008.*
[10] *mg/kg of grape berry fresh weight.*
[11] *Average values from two vintages (2006/2007 and 2007/2008).*
[12] *Values expressed as mg/l.*
[13] *Average values from vintage 2010.*

Table 2.
General physiochemical and phenolic composition of several international and native red grape varieties cultivated under different geographical locations.

years, several studies have been published that present results on the adaptability of the various varieties in the different regions with conditions for grape wine production. However, some of these regions have a warm climate, characteristic of tropical regions. In these regions, the minimum temperature is not sufficiently low to induce natural vegetative repose in the vines. Oliveira et al. [71] reported chemical characteristics of grapes Syrah grown in a Brazilian tropical semiarid region (Pernambuco State) during four growing seasons (two calendar years, 2016 and 2017). According to these authors, in the semiarid region considered, the interaction between the Syrah grape and the climatic conditions in each harvest season (combined also with the different rootstocks) determines the composition of the grapes. For example, grapes harvested in July of the first harvest season, where the temperatures are lower, showed higher total acidity, total monomeric anthocyanins, and total tannins in seeds, than grapes harvested in December from the second harvest season, characterized warmer temperatures. Also, Stefanello et al. [72] studied the potential of Alicante Bouschet variety in the Campanha Gaúcha region, southern Brazil, between 2013 and 2017. At the same time, yield and chemical composition of the grape must subjected to nitrogen application without irrigation, followed by irrigation and via fertigation were evaluated. These authors concluded a good adaptation of this French variety to environmental conditions of Southern Brazil and at same time the grapevines grown in control soil without Nitrogen fertilization had the highest values of total soluble solids in the must in all the crop seasons considered. In 2011, Borghezan et al. [73] also reported that Cabernet Sauvignon, Merlot, and Sauvignon Blanc produced grapes with high quality, being suitable for cultivation in São Joaquim, Santa Catarina State (Brazil).

Chile and Argentine show in general red and white grape varieties from different origins well adapted to the different terroirs in the diverse wine regions. Among Chilean varieties, Carignan Noir, a variety of Spanish origin (Aragón region), has had a major resurgence due to its rediscovered wine quality potential. Martínez-Gil et al. [74] reported results for this red grape variety by the characterization of phenolic composition of grapes grown in different locations from the Maule Valley. The data obtained show high enological and viticultural attributes for grape growers of this cultivar inducing differentiable attributes in terms of grape composition. For Argentina, Malbec is the most important variety. This is a red grape variety originated from France; however, Argentina has the highest acreage of vineyards of this variety (representing around 77% of the world production) being emblematic for Argentina's winemaking industry [75]. Several, authors describe that this variety requires an intermediate to warm climate, low rainfall, and with solar potential. These conditions are found in several Argentinian wine regions producing Malbec grapes and wines with a high quality [76, 77].

Finally, it is important to considerate the adaptability of different international grapes varieties in one of the biggest and dynamic wine country as is the case of China. This country has had a great success in the grape and wine production with an extraordinary development. Most grapes are from red varieties (around 80%), while the white varieties represent only 20% [78]. For red varieties, Cabernet Sauvignon, Merlot, and Cabernet Gernischt have a great adaptability to the different Chinese wine regions with diverse climatic and agronomic conditions. The white grapes, Italian Riesling, Chardonnay, and Riesling, are the most cultivated varieties [79, 80]. In last years, several studies were carried out to analyze the adaptability of the different *V. vinifera* varieties in the different Chinese wine regions under diverse climatic and soil conditions combined with several agronomic practices. The results show a good adaptability for most of the cultivars maintaining all its specific potentialities but affected by the environmental conditions [81, 82].

4. Influence of different environmental conditions on wine composition

Apart from grape composition of the different international varieties cultivated under different environmental conditions, several studies have been carried out on the composition of the wines produced and their sensory profile. However, although international varieties are in general well adapted to specific environmental and production conditions, the wines from native varieties are shown to possess at least equal potential concerning the quality. Žurga et al. [83] reported comparative data from Croatian wines made from native (Plavac Mali and Teran) and non-native grape varieties (Merlot and Cabernet Sauvignon), both cultivated in Croatian coastal regions. According to these authors, wine produced from Plavac native grape variety had the highest total phenolic and (+)-catechin content, while Merlot and Cabernet Sauvignon wines had higher flavonol content. According to these authors, there are distinct genetic potentials between Croatian autochthonous and no autochthonous (specifically for Merlot and Cabernet Sauvignon) varieties. Additionally, several authors reported data from Merlot and Cabernet Sauvignon wines elaborated in Serbian Fruška gora wine region (Northern Serbia). Similar total phenols (ranged between 1460 and 1560 mg GAE/l) and total anthocyanins (ranged between 329 and 319 mg malvidin-3-glucoside equivalents/l) content were found between these wines [84–87]. Also, Vujović et al. [66] studied the impact of three Merlot clones uses on wine quality under Central Serbia wine region conditions (Grocka wine region). These authors found total phenols values being between 1100 and 1500 mg GAE/l. Sredojević [87] compared young wines produced from Cabernet Sauvignon variety and Serbian native varieties (Prokupac, Crna Tamjanika, Plovdina, Smederevka, and Kreaca). This author found higher values of total phenols and total anthocyanins for the wine produced from Cabernet Sauvignon. Diverse results were found by Pantelić et al. [86], which revealed that the red wine from Serbian variety Prokupac grown in Central Serbia showed higher values of total phenols, (+)-catechin, and (−)-epicatechin compared with the wines from Cabernet Sauvignon, Merlot, Cabernet Franc, Syrah, and Pinot Noir produced in the same wine region. The differences in the conditions of the terroirs in these last two studies had significant impact on the values of the abovementioned parameters.

In Spain, namely in Galicia region (Northwestern Spain), where wine production in mainly focused on white wines, several monovarietal white wines produced from native (Albariño, Branco Lexitimo, Caiño Blanco, Godello, Loureiro, Torrontes, and Treixadura) and non-native (Chardonnay, Gewürztraminer, Pinot Blanc, Pinot Gris, Riesling, and Sauvignon Blanc) varieties were study [88]. This study reported that the wines produced from native varieties had a clear differentiation by their phenolic composition. Specifically, Caiño and Treixadura wines showed the highest total phenolic content, while Riesling presented the highest values among wines produced from non-native varieties (although with values below those observed for native varieties). In China, Li et al. [89] compared the phenolic and chromatic characteristics of red wines produced from native (*Vitis amurensis* and its hybrids, and *Vitis davidii*) and several international (Pinot Noir, Marselan, Cabernet Sauvignon, and Syrah) varieties. For these authors, there are specific phenolic compounds that could be recognized as phenolic fingerprints of wines, which could explain their chromatic differences. For example, wines produced from native Chinese varieties had relatively higher blue % value and lower red % value compared with wines obtained from international varieties.

The Cabernet Sauvignon grape represents about 7% of the vineyard area in Argentina. Recently, Muñoz et al. [31] studied the chemical and sensory characteristics of the Cabernet wines in different geographical regions during 2018 and 2019 vintages. They concluded that this variety shows a good adaptability to the different Argentina wine regions; however, the selection of the plant material (different clones) combined with the terroir determines the quality attributes of the wines produced.

Table 3 shows results for some wine phenolic compounds for several red wine produced from different international varieties (Cabernet Sauvignon, Merlot, Syrah, Pinot Noir, and Malbec) cultivated in the main South American wine producers. By the data show in this table, it is clear a great diversify of the wine phenolic content. For total anthocyanins, a high variation of values is found, varying between 69.4 and 310 and 681.8 mg/l, for Cabernet Sauvignon wines produced in Brazil and Argentina, respectively. In addition, Malbec wines from Argentina show the highest total average anthocyanin content [90–92]. Recently, also in Brazil, a few authors reported for Merlot wines produced in high altitude regions of the State of Santa Catarina, high concentrations of hydroxycinnamic acids and several volatile compounds, such as phenylethyl acetate, ethyl cinnamate, and γ-lactone, which contribute to the aromas of coconut, peach, roses, honey, and red fruits. All this content observed was comparable with the results obtained for red wines produced from this grape variety from other regions and countries [94]. Faustino et al. [95] reported a comparative analysis of phenolic composition of selected American, Chilean, and Canadian Merlot wines. These authors concluded that the different climatic regions have and important role of Merlot wines phenolic composition. Thus, Chilean Merlot wines tended to have higher flavonoid content, while Canadian wines showed intermediate values, and American wine had the lowest flavonoid values. The results obtained also demonstrated that moderate average temperatures provide Merlot wines with high phenolic content.

For Syrah wines obtained from grapes cultivated in two regions with different altitudes in Northeast Brazil and during two vintages (2014 and 2015), it was demonstrated that chemical composition of wines was influenced by altitude. In that case, wines obtained from higher altitude (1100 m) region showed the highest phenolic composition. However, for sensory profile, floral, herbaceous, fruity, and empyreumatic aromatic attributes were obtained in Syrah wines from the 350 m altitude region [96]. Also, a study published by Fushing and colleagues [76] reported the relationships between chemical and sensory characteristics of Malbec wines in connection with their regions of production. According to these authors, there is a more marked regionality in Argentinian Malbec wines compared with Californian Malbec wines.

In addition, King et al. [97] also reported for Malbec wines from Mendoza region (Argentina) that generally they show more ripe fruit, sweetness, and higher alcohol levels, while the Californian Malbec wines have more synthetic fruit and citrus aromas and bitter taste. Compositional differences between the wines from these two countries were related more to altitude than precipitation and growing degree days. In fact, the viticultural sites in Mendoza, Argentina, are located at much higher elevations (on average, 1103 ± 133 m above sea level), than those in California (on average, 190 ± 200 m above sea level).

Table 4 summarizes several results obtained for general chemical analytical parameters and phenolic composition of different monovarietal red wines produced from Malbec, Bonarda, Cabernet Sauvignon, Merlot, Syrah, and Tempranillo varieties in Mendoza, Argentina. The results found by Fanzone et al. [92] in this region are indicative of the polyphenolic richness of Malbec compared with the other red varieties and their potential to produce quality wines. Titratable acidity varied from 4.4 to 6.8 g/l, pH from 3.60 to 3.84, and ethanol content from 13.0 to 15.2%. These parameters influenced the sensory quality and microbiological stability of the wine.

DOI: http://dx.doi.org/10.5772/intechopen.108204

Varieties/countries		Phenolic compounds						
		Total anthocyanins	Cinnamic acids		Flavan-3-ols		Benzoic acids	Stilbene
			p-coumaric acid[2]	Caffeic acid[2]	(+)-Catechin[2]	(−)-Epicatechin[2]	Gallic acid[2]	Trans-resveratrol[2]
Cabernet Sauvignon	Brazil	69.4–310[1]	8.02	4.62	76.46	25.64	37.92	3.34
	Argentina	681.8[3]	4.81	2.74	116.18	24.03	62.71	1.56
	Chile	246.0[2]	4.98	4.95	125.61	29.65	51.06	2.10
Merlot	Brazil	89.1–310[1]	10.73	3.61	64.39	25.88	32.48	3.55
	Argentina	644.1[3]	7.56	3.03	85.57	20.12	63.24	2.66
	Chile	109–121[1]	7.51	4.87	149.14	30.69	69.87	4.21
Syrah	Brazil	92.1–386[1]	6.97	5.42	121.17	36,39	52.94	2.83
	Argentina	301.4[3]	5.58	4.21	83.36	22.73	55.33	3.41
	Chile	74.5–199[3]	6.88	4.91	86.94	39.51	44.01	3.14
Pinot Noir	Brazil	100.36[4]	6.17	3.80	59.15	28.40	13.88	3.82
	Argentina	—	4.40	4.61	123.02	33.99	27.64	2.90
	Chile	—	4.97	4.93	93.23	41.72	41.96	4.30
Malbec	Brazil	—	—	—	—	—	—	—
	Argentina	1044.5[3]	6.91	345	88.62	19.75	41.75	2.36
	Chile	—	7.21	4.30	67.84	44.53	33.49	2.07

[1]*De Andrade et al. [90].*
[2]*Granato et al. [91].*
[3]*Fanzone et al. [92].*
[4]*Nascimento et al. [93].*

Table 3.
Phenolic composition (mg/l) of red wines produced from several international varieties in three South American countries.

Parameters/compounds	Wines/varieties					
	Malbec	Bonarda	Cabernet Sauvignon	Merlot	Syrah	Tempranillo
Titratable acidity (g/l tartaric acid)	6.8 ± 0.3	5.3 ± 0.2	6.7 ± 0.3	5.8 ± 0.4	4.4 ± 0.2	5.6 ± 0.3
pH	3.6 ± 0.06	3.8 ± 0.04	3.73 ± 0.09	3.73 ± 0.09	3.8 ± 0.06	3.84 ± 0.07
Ethanol (%, v/v)	15.2 ± 0.4	13.0 ± 0.3	14.5 ± 0.3	14.5 ± 0.1	13.6 ± 0.2	13.8 ± 0.2
Total phenols (mg/l GAE)	4203.2 ± 412.8	3372.1 ± 453.0	3377.6 ± 369.6	3447.5 ± 372.3	1585.6 ± 50.6	3137.4 ± 152.9
Total anthocyanins (mg/l)[1]	1044.5 ± 88.2	739.8 ± 55.0	681.8 ± 100.8	644.1 ± 37.6	301.4 ± 18.9	717.6 ± 41.9
Proanthocyanidins (mg/l catechin)	5013.0 ± 507.2	3925.4 ± 556.3	3860.7 ± 439.6	4439.9 ± 498.5	1922.6 ± 160.1	3200.7 ± 250.9
Malvidin-3-glucoside (mg/l)[1]	257.0 ± 8.0	146.2 ± 6.4	141.3 ± 12.2	119.3 ± 10.6	73.7 ± 2.2	162.0 ± 7.6
Peonidin-3-glucoside (mg/l)[1]	40.3 ± 2.4	16.6 ± 1.2	14.9 ± 1.3	24.3 ± 2.3	14.3 ± 1.0	17.0 ± 1.0
Petunidin-3-glucoside (mg/l)[1]	61.8 ± 2.3	28.7 ± 2.0	14.4 ± 0.9	23.2 ± 1.9	6.0 ± 0.3	35.0 ± 0.7
Cyanidin-3-glucoside (mg/l)[1]	5.48 ± 0.28	2.44 ± 0.20	2.10 ± 0.20	2.46 ± 0.17	0.66 ± 0.04	2.42 ± 0.15
Delphinidin-3-glucoside (mg/l)[1]	41.2 ± 2.4	18.9 ± 1.5	10.1 ± 1.0	14.5 ± 1.1	2.7 ± 0.1	26.5 ± 1.5
Total glucosylated anthocyanins (mg/l)[1]	405.8 ± 11.8	212.8 ± 10.1	182.8 ± 14.8	183.8 ± 10.8	97.4 ± 3.3	242.9 ± 10.8
trans-Resveratrol-3-glucoside (mg/l)	9.2 ± 1.0	3.2 ± 0.4	2.1 ± 0.3	4.1 ± 0.6	2.2 ± 0.4	1.9 ± 0.1
(+)-Catechin (mg/l)	52.7 ± 6.5	58.7 ± 5.4	52.2 ± 3.7	44.0 ± 3.5	28.5 ± 5.7	19.9 ± 0.3
(−)-Epicatechin (mg/l)	21.8 ± 3.3	34.8 ± 3.0	26.0 ± 1.4	31.7 ± 1.9	14.9 ± 3.6	12.3 ± 1.4

[1]*Values expressed as equivalents of malvidin-3-monoglucoside.*

Table 4.
General chemical parameters and individual phenolic composition of several monovarietal red wines produced from Malbec, Bonarda, Cabernet Sauvignon, Merlot, Syrah, and Tempranillo in Mendoza region, Argentina (Fanzone et al. [92]).

Malbec wines presented higher acidity, lower pH, and higher ethanol content than the other varieties. Total phenols ranged from 1585.6 to 4203.2 mg/l, and Malbec wines showed higher phenolic levels than others, while Syrah wines had the lowest phenolic content. Malbec wines had also the highest content of total anthocyanins and proanthocyanidins.

In Europe, the adaptability of international grape varieties outside from their terroir of origin creates wines with diverse chemical and sensory profiles. Cosme et al. [50] studied the tannin profile of several monovarietal wines obtained from different Portuguese native (Touriga Nacional, Trincadeira, and Castelão) and no native (Syrah and Cabernet Sauvignon) red grape varieties cultivated in Lisbon wine region during two vintages (2004 and 2005). These authors reported similar tannin profiles in each vintage for Trincadeira and Cabernet Sauvignon wines. Monagas et al. [59] reported data about the wine flavan-3-olis composition for Graciano, Tempranillo, and Cabernet Sauvignon wines elaborated under the same conditions and obtained from grapes cultivated in the same geographical area (Navarra, Spain) and also with same technological maturity. These authors described that among the three wines, similar skins' proanthocyanidin content was obtained. However, Cabernet Sauvignon wines showed a lower mean degree of polymerization of proanthocyanidins than Tempranillo wines.

It is well known that volatile composition is crucial to the quality of wines due to their influence on the aroma profile. For Chardonnay wines produced in Spanish Castilla-La Mancha wine region, a few studies reported that these white wines in general show a high content of several terpenes, such as geraniol, citronellol, and linalool, which provide citric and floral aromas [98, 99]. These contents are even higher than those found in other white wines produced from Spanish native grape varieties [100]. A comparative analysis of the sensory flavor characteristics between monovarietal white wines produced from different native grape varieties cultivated in Turkey and Chardonnay and Semillon wines was studied by Elmaci et al. [101]. The wines produced from the two international white varieties showed specific aroma and flavor characteristics, where notes of green plum for Semillon and banana and tobacco for Chardonnay wines were reported as specific sensory characteristics compared with white wines obtained from native Turkish varieties.

Finally, the continuous increase in global temperatures is leading to the appearance of new wine regions, outside of the traditional regions in the "old world" (Italy, France, Spain, Portugal, among other European countries) and "new world" (Chile, Argentina, South African, Australia, or United States). Thus, new wine-producing countries from different areas, such as the northern parts of Europe, are emerging. In this case, international grapes varieties have been introduced besides the use of a few hybrid cultivars varieties highly resistant to the specific climatic conditions (i.e., frost) and tolerant to oidium, downy mildew, and to some extent also to *Botrytis*.

Recently, Garrido-Bañuelo et al. [102] compared several Swedish white wines produced from Solaris variety (a hybrid cultivar) with different white wines produced from Sauvignon Blanc (from New Zealand and France), Chardonnay (from France), and Chenin blanc (from South Africa). In general, the differences in taste and mouthfeel were more obvious between Solaris wines and the other white wines produced from the other international varieties independently of the country origin. In England, the vineyard area in 2013 had has grown with Pinot Noir and Chardonnay representing 22 and 23% of vineyard area, respectively. The wines produced from these two international cultivars represent a substantial value of the total wine produced in this country [103].

Lastly, it is important to note that in general, wine consumers attribute a high value to the aspects related to typicality and consequently to native grape varieties to produce unique wines. However, this evidence runs parallel to another tendency that moves in the contrary trend, that is, the emergence of "international" tastes among the different consumers, especially in no traditional wine countries producers. Boncinelli et al. [104] reported that the initial wine consumers' preference for wines with only native grape varieties is contradicted in the post-taste evaluation, where wines containing between 10 and 20% of international grape varieties are in general more valorized than wines containing 100% of native varieties.

5. Final remarks

In this chapter, we focused on the adaptation of most important international grape varieties, originating from traditional wine countries, in different world wine regions characterized by heterogeneous environmental conditions and specific "terroirs." The adaptability is assessed through the available data on most important productivity parameters, as well as, through grape and wine composition, and sensory profile.

Grape and wine composition of the same vine varieties, but grown in different wine regions of the world, depends on climate factors, such as sunlight exposition, temperature, rainfall, as well as soil type and composition, degree of grape ripeness, and of course on cultural practices involved in grape processing and wine production. Winegrowers had traditionally selected and cultivated foreign grape varieties that best match their specific climate and soil conditions. However, the impact of climate changes on vitivinicultural sector is already obvious in various aspects, causing the experts from different fields focus on research and experimental activities about the adaptive capacity of the most economically important grape varieties to these global changes in different wine regions. Furthermore, the continuous increase in global temperatures resulted in the appearance of new wine regions as well as in the further dispersion of various grape varieties in areas where viticulture was, until recently, very limited. Thus, new wine-producing regions, such as northern parts of Europe and China, and areas in North and South America are emerging. These novel wine regions are oriented on the cultivation of both established grape varieties and relatively new hybrid varieties highly resistant to specific climate conditions (i.e., frost) and tolerant to oidium, downy mildew, and to some extent also to *Botrytis*. In addition, other important challenges are certainly the clonal selection process of these varieties ensuring the clones with best properties in the specific environments. Thus, the scientific knowledge about the adaptability of different international grape varieties in diverse environments is of great importance for winegrowers worldwide.

Conflict of interest

The authors declare no conflict of interest.

Author details

Tatiane Otto[1], Renato Botelho[1], Luiz Biasi[2], Uroš Miljić[3], Ana C. Correia[4] and António M. Jordão[4,5]*

1 State University of Midwestern of Paraná, Guarapuava, Brazil

2 Federal University of Paraná, Curitiba, Brazil

3 Faculty of Technology, University of Novi Sad, Novi Sad, Serbia

4 Agrarian Higher School, Polytechnic Institute of Viseu, Viseu, Portugal

5 Chemistry Research Centre of Vila Real (CQ-VR), Vila Real, Portugal

*Address all correspondence to: antoniojordao@esav.ipv.pt

References

[1] Vitis International Variety Catalogue (VIVC). Germany: Institute of Grapevine Breeding - Gellweilerhof, Julius Kühn-Institut; 2022. [Internet] Available from: http://www.vivc.de [Accessed: March 1, 2022]

[2] Olmo HP. The potential role of (*vinifera*. X *rotundifolia*) hybrids in grape variety improvement. Experientia. 1986;**42**:921-926

[3] Wen J, Soejima A, Meng Y. Phylogeny of *Vitaceae* based on the nuclear GAI1 gene sequences. Canadian Journal of Botany. 2007;**85**:731-745. DOI: 10.1139/B07-071

[4] Dehan C, editor. Cépages et vignobles de France—Tome 1: Les vignes américaines. 2nd ed. France: Paul Déhan; 1988. 553 p Pierre Galet Ed

[5] Maul E, Topfer R. Vitis international variety catalogue (VIVC): A cultivar database referenced by genetic profiles and morphology. BIO Web of Conferences. 2015;**5**:01009. DOI: 10.1051/bioconf/20150501009

[6] Lacombe T, editor. Contribution à l'étude de l'histoire évolutive de la vigne cultivée (*Vitis vinifera* L.) par l'analyse de la diversité génétique neutre et de gènes d'intérêt. France: Montpellier; 2012. 328 p

[7] International Organisation of Vine and Wine 2022. State of the world vitivinicultural sector in 2021. [Internet] Available from: https://www.oiv.int/. [Accessed: April 4, 2022]

[8] Lecat B, Amspacher W, Higgins L, Ferrara AL, Wolf M. Wine sector: Definitions and nuances from global to country analysis—A comparison between Old World, New World, and emerging wine countries from 2005 to current. In: Santini C, Cavicchi A, editors. Case Studies in the Wine Industry. In the Consumer Science and Strategic Marketing Series Food Science, Technology and Nutrition. Vol. 2019. Amsterdam: Elsevier Publisher. pp. 7-32. DOI: 10.1016/B978-0-08-100944-4.00002-1

[9] International Organisation of Vine and Wine (OIV). Paris, France; 2021. [Internet]. Available from: https://www.oiv.int/ [Accessed: January 5, 2022]

[10] Schultz HR. Climate change and viticulture: A European perspective on climatology, carbon dioxide and UV-B effects. Australian Journal of Grape and Wine Research. 2000;**6**:2-12. DOI: 10.1111/j.17550238.2000.tb00156.x

[11] Koufos GC, Mavromatis T, Koundouras S, Jones GV. Adaptive capacity of wine grape varieties cultivated in Greece to climate change: Current trends and future projections. OENO One. 2020;**54**:1201-1219. DOI: 10.20870/oenoone.2020.54.4.3129

[12] Suter B, Destrac IA, Gowdy M, Dai Z, Van Leeuwen C. Adapting wine grape ripening to global change requires a multi-trait approach. Frontiers in Plant Science. 2021;**12**:624867. DOI: 10.3389/fpls.2021.624867

[13] Anderson K, Aryal N. Where in the world are various wine grape varieties grown? Evidence from a new database. Bulletin de l'OIV. 2013;**86**(992-993-994):461-484

[14] Clímaco P, Cunha JP. Exigences thermiques de quelques cépages cultives dans la region de l'Estremadura. Ciência e Técnica Vitivinícola. 2006;**21**:67-74

[15] Ó-Marques J, Reguinga R, Laureano O, Ricardo-da-Silva JM. Changes in grape seed, skin, and pulp condensed tannins during berry ripening: Effect of fruit pruning. Ciência e Técnica Vitivinícola. 2005;**20**:35-42

[16] Lorrain B, Chira K, Teissedre PL. Phenolic composition of merlot and cabernet-sauvignon grapes from Bourdeaux vineyard for the 2009-vintage: Comparison to 2006, 2007 and 2008 vintages. Food Chemistry. 2011;**126**:1991-1999. DOI: 10.1016/j.foodchem.2010.12.062

[17] Costa E, Cosme F, Rivero-Pérez MD, Jordão AM, González-SanJosé ML. Influence of wine region provenance on phenolic composition, antioxidant capacity and radical scavenger activity of traditional Portuguese red grape varieties. European Food Research and Technology. 2015;**241**:61-73. DOI: 10.1007/s00217-015 2434-x

[18] Blancquaert EH, Oberholster A, Ricardo-da-Silva JM, Deloire AJ. Grape flavonoid evolution and composition under altered light and temperature conditions in cabernet sauvignon (*Vitis vinifera* L.). Frontiers in Plant Science. 2019;**10**:1062. DOI: 10.3389/fpls.2019.01062

[19] Costa E, Cosme F, Jordão AM, Mendes-Faia A. Anthocyanin profile and antioxidant activity from 24 grape varieties cultivated in two Portuguese wine regions. Journal International des Sciences de la Vigne et du Vin. 2014;**48**:51-62. DOI: 10.20870/oeno-one.2014.48.1.1661

[20] Domec JC, Johnson DM. Does homeostasis or disturbance of homeostasis in minimum leaf water potential explain the isohydric versus anisohydric behavior of *Vitis vinifera* L. cultivars? Tree Physiology. 2012;**32**:245-248. DOI: 10.1093/treephys/tps013

[21] Wang R, Sun Q, Chang Q. Soil types effect on grape and wine composition in Helan Mountain area of Ningxia. PLoS One. 2015;**10**(2):e0116690. DOI: 10.1371/journal.pone.0116690

[22] Stefanos K. Environmental and viticultural effects on grape composition and wine sensory properties. Elements. 2018;**14**(3):173-178. DOI: 10.2138/gselements.14.3.173

[23] Obreque-Slier E, López-Solís R, Castro-Ulloa L, Romero-Díaz C, Peña-Neira A. Phenolic composition and physicochemical parameters of Carménère, cabernet sauvignon, merlot, and cabernet franc grape seeds (*Vitis vinifera* L.) during ripening. LWT—Food Science and Technology. 2012;**48**:134-141. DOI: 10.1016/j.lwt.2012.02.007

[24] Wolkovich EM, Garcia de Cortazar-Atauri I, Morales-Castilla I, Nicholas KA, Lacombe T. From pinot to xinomavro in the world's future wine-growing regions. Nature Climate Change. 2018;**8**:29-37. DOI: 10.1038/s41558-017-0016-6

[25] Drori E, Munitz S, Pinkus A, Stanevsky M, Netzer Y. The effect of irrigation-initiation timing on the phenolic composition and overall quality of cabernet sauvignon wines grown in a semi-arid climate. Food. 2022;**11**(5):770. DOI: 10.3390/foods11050770

[26] International Organisation of Vine and Wine. OIV Focus 2017/Vine varieties distribution in the world. Paris, France: OIV Publisher; 2017 [Internet]. Available from: https://www.oiv.int/ [Accessed: March 25, 2022]

[27] Song GC. Grape production in the Republic of Korea. In:

Papademetriou MK, Dent FJ, editors. Grape Production in the Asia-Pacific Region. 1st ed. Bangkok, Thailand: FAO; 2001. pp. 53-60

[28] Galet P. Dictionnaire encyclopédique des cépages et de leurs synonymes. In: Ed. Libre and Solidaire. Paris, France: Libre & Solidaire Publisher; 2015. p. 1199

[29] Nicotra AB, Davidson A. Adaptive phenotypic plasticity and plant water use. Functional Plant Biology. 2010;**37**:117-127. DOI: 10.1071/FP09139

[30] Dal Santo S, Fasoli M, Negri S, D'Incà E, Vicenzi N, Guzzo F, et al. Plasticity of the berry ripening program in a white grape variety. Frontiers in Plant Science. 2016;**7**:970. DOI: 10.3389/fpls.2016.00970

[31] Muñoz F, Urvieta R, Buscema F, Rasse M, Fontana A, Berli F. Phenolic characterization of cabernet sauvignon wines from different geographical indications of Mendoza, Argentina: Effects of plant material and environment. Frontiers in Sustainable Food Systems. 2021;**5**:700642. DOI: 10.3389/fsufs.2021.700642

[32] Anesi A, Stocchero M, Dal Santo S, Commisso M, Zenoni S, Ceoldo S, et al. Towards a scientific interpretation of the terroir concept: Plasticity of the grape berry metabolome. BMC Plant Biology. 2015;**15**:1-17. DOI: 10.1186/s12870-015-0584-4

[33] Marlowe B, Lee S. Conceptualizing terroir wine tourism. Tourism Review International. 2018;**22**(2):143-151. DOI: 10.3727/154427218X1531 92863.72298

[34] Magalhães N, editor. Tratado de Viticultura: A Videira, a Vinha e o Terroir. Lisboa, Portugal: Esfera Poética Publisher; 2015. p. 605

[35] Carbonneau A. Ecophysiologie de la vigne et terroir. In: Terroir, Zonazione, Viticoltura. Trattato Internazionale; Centreville, VA, USA: Phytoline; 2003. pp. 61-102

[36] Van Leeuwen C, Friant P, Choné X, Tregoat O, Koundouras S, Dubordieu D. Influence of climate, soil, and cultivar on terroir. American Journal of Enology and Viticulture. 2004;**55**:207-217

[37] Austin ME, Bondari K. A study of cultural and environmental-factors on the yield of Vitis-Rotundifolia. Scientia Horticulturae. 1988;**34**:219-227

[38] Fraga H, Pinto JG, Santos JA. Climate change projections for chilling and heat forcing conditions in European vineyards and olive orchards: A multi-model assessment. Climatic Change. 2019;**152**:179-193. DOI: 10.1007/s10584-018-2337-5

[39] Riou C, Carbonneau A, Becker N, Caló A, Costacurta A, Castro R, et al., editors. Le Determinisme climatique de la maturation du raisin: application au zonage de la teneur em sucre dans la communauté européenne. Luxembourg: Office des publications officielles des communautés européennes; 1994. p. 319

[40] Peacock WL, Rolston DE, Aljibury FK, Rauschkolb RS. Evaluating drip, flood, and sprinkler irrigation of wine grapes. American Journal of Enology and Viticulture. 1977;**28**:193-195

[41] Butt TM, Copping LG. Fungal biological control agents. Pesticide Outlook. 2000;**11**:186-191

[42] Molitor D, Baron N, Sauerwein T, André CM, Kicherer A, Döring J, et al. Postponing first shoot topping reduces grape cluster compactness and delays bunch rot epidemic. American Journal of Enology and Viticulture. 2015;**66**:164-176

[43] De la Fuente LM. Use of hybrids in viticulture. A challenge for the OIV. OENO One. 2018;**52**(3):231-234. DOI: 10.20870/oenoone.2018.52.3.2312

[44] Karvonen J. Does soil temperature restrict outdoor viticulture in southern Finland? Vitis. 2013;**52**(4):207-208. DOI: 10.5073/vitis.2013.52.207-208

[45] Caliari V, Burin VM, Rosier JP, Bordignonluiz MT. Aromatic profile of Brazilian sparkling wines produced with classical and innovative grape varieties. Food Research International. 2014;**62**:965-973. DOI: 10.1016/j.foodres.2014.05.013

[46] De Bem B, Brighenti E, Bonin BF, Allembrandt R, Araujo L, Brighenti AF, et al. Downy mildew intensity in tolerant grapes varieties in highlands of southern Brazil. BIO Web of Conferences. 2016;**7**:01015. DOI: 10.1051/bioconf/20160701015

[47] Tello J, Cordero-Bueso G, Aporta I, Cabellos JM, Arroyo T. Genetic diversity in commercial wineries: Effects of the farming system and vinification management on wine yeasts. Journal of Applied Microbiology. 2012;**112**:302-315. DOI: 10.1111/j.1365-2672.2011.05202.x

[48] Jones GV. Climate and terroir: Impacts of climate variability and change on wine. In: Macqueen RW, Meinert LD, editors. Fine Wine and Terroir—The Geoscience Perspective. Canada: Geoscience Canada Reprint; 2006. p. 246

[49] Costa E, Da Silva JF, Cosme F, Jordão AM. Adaptability of some French red grape varieties cultivated at two different Portuguese terroirs: Comparative analysis with two Portuguese red grape varieties using physicochemical and phenolic parameters. Food Research International. 2015;**78**:302-312. DOI: 10.1016/j.foodres.2015.09.029

[50] Cosme F, Ricardo-Da-Silva JM, Laureano O. Tannin profiles of *Vitis vinifera* L. cv. Red grapes growing in Lisbon and from their monovarietal wines. Food Chemistry. 2009;**112**:197-204. DOI: 10.1016/j.foodchem.2008.05.058

[51] Jackson DL, Lombard PB. Environmental and management practices affecting grape composition and wine quality review. American Journal of Enology and Viticulture. 1993;**4**:409-430

[52] Gordillo B, Rodríguez-Pulido FJ, Mateus N, Escudero-Gilete ML, González-Miret ML, Heredia FJ, et al. Application of LC–MS and tristimulus colorimetry to assess the ageing aptitude of Syrah wine in the Condado de Huelva D.O. (Spain), a typical warm climate region. Analytica Chimica Acta. 2012;**732**:162-171. DOI: 10.1016/j.aca.2012.03.051

[53] González-Neves G, Franco J, Barreiro L, Gil G, Moutounet M, Carbonneau A. Varietal differentiation of Tannat, cabernet-sauvignon and merlot grapes and wines according to their anthocyanic composition. European Food Research and Technology. 2007;**225**:111-117. DOI: 10.1007/s00217-006-0388-8

[54] Flamini R, Mattivi F, Rosso M, Arapitsas P, Bavaresco L. Advanced knowledge of three important classes of grape phenolics: Anthocyanins, stilbenes and flavonols. International Journal of Molecular Sciences. 2013;**14**:19651-19669. DOI: 10.3390/ijms141019651

[55] Sikuten I, Stambuk P, Tomaz I, Marchal C, Kontic JK, Lacombe T, et al. Discrimination of genetic and geographical groups of grape varieties (*Vitis vinifera* L.) based on their polyphenolic profiles. Journal of Food Composition and

Analysis. 2021;**102**:104062. DOI: 10.1016/j.jfca.2021.104062

[56] Castellarin S, Matthews M, Gaspero G, Gambetta G. Water deficits accelerate ripening and induce changes in gene expression regulating flavonoid biosynthesis in grape berries. Planta. 2007;**227**:101-112. DOI: 10.1007/s00425-007-0598-8

[57] Deluc LG, Quilici DR, Decendit A, Grimplet J, Wheatley MD, Schlauch KA, et al. Water deficit alters differentially metabolic pathways affecting important flavor and quality traits in grape berries of cabernet sauvignon and chardonnay. BMC Genomics. 2009;**10**:212. DOI: 10.1186/1471-2164-10-212

[58] Yang J, Martinson TE, Liu RH. Phytochemical profiles and antioxidant activities of wine grapes. Food Chemistry. 2009;**116**:332-339. DOI: 10.1016/j.foodchem.2009.02.021

[59] Monagas M, Gómez-Cordovés C, Bartolomé B, Laureano O, Ricardo da Silva JM. Monomeric, oligomeric, and polymeric flavan-3-ol composition of wines and grapes from *Vitis vinifera* L. Cv. Graciano, Tempranillo, and Cabernet Sauvignon. Journal of Agricultural and Food Chemistry. 2003;**51**(22):6475-6481. DOI: 10.1021/jf030325

[60] García E, Chacón JL, Martínez J, Izquierdo PM. Changes in volatile compounds during ripening in grapes of Airén, Macabeo and chardonnay white varieties grown in La Mancha region (Spain). Food Science and Technology International. 2003;**9**(1):33-41. DOI: 10.1177/1082013203009001006

[61] Mitić M, Souquet JM, Obradović M, Mitić S. Phytochemical profiles and antioxidant activities of Serbian table and wine grapes. Food Science Biotechnology. 2012;**21**(6):1619-1626. DOI: 10.1007/s10068-012-0215-x

[62] Pantelić M, Dabić Zagorac D, Natić M, Gašić U, Jović S, Vujović D, et al. Impact of clonal variability on phenolics and radical scavenging activity of grapes and wines: A study on the recently developed merlot and cabernet franc clones (*Vitis vinifera* L.). PLoS One. 2016;**11**(10):0163823. DOI: 10.1371/journal.pone.0163823

[63] Pantelić M, Dabić Zagorac D, Davidović S, Todić S, Bešlić Z, Gašić U, et al. Identification and quantification of phenolic compounds in berry skin, pulp, and seeds in 13 grapevine varieties grown in Serbia. Food Chemistry. 2016;**211**:243-252. DOI: 10.1016/j.foodchem.2016.05.051

[64] Ivanišević D, Kalajdžić M, Drenjančević M, Puškaš V, Korać N. The impact of cluster thinning and leaf removal timing on the grape quality and concentration of monomeric anthocyanins in cabernet sauvignon and probus (*Vitis vinifera* L.) wines. OENO One. 2020;**1**:63-74. DOI. DOI: 10.20870/oeno-one.2020.54.1.2505

[65] Cindrić P, Korać N, Kovač V. Cabernet sauvignon, cabernet franc and merlot in viticultural area of Fruska gora. Poljoprivreda. 1996;**45**:35-40

[66] Vujović D, Žunić D, Popović B, Pantelić M, Popović-Djordjević J. Agrobiological and wine quality traits of *Vitis Vinifera* cv. Merlot clones selected in Serbia. Journal International Des Sciences De La Vigne Et Du Vin. 2015;**49**(4):267-274. DOI. DOI: 10.20870/oeno-one.2015.49.4.1

[67] Oliveira JB, Faria DL, Duarte DF, Egipto R, Laureano O, Castro R, et al. Effect of the harvest season on phenolic composition and oenological parameters of grapes and wines cv. 'Touriga Nacional' (*Vitis vinifera* L.) produced under tropical semi-arid climate, in the state of Pernambuco, Brazil. Ciência e

Técnica Vitivinícola. 2018;**33**(2):145-166. DOI: 10.1051/ctv/20183302145

[68] Oliveira JB, Faria DL, Duarte DF, Egipto R, Laureano O, Castro R, et al. Climate effects on physicochemical composition of Syrah grapes at low and high-altitude sites from tropical grown regions of Brazil. Food Research International. 2019;**121**:870-879. DOI: 10.1016/j.foodres.2019.01.011

[69] Hunter JJ, Volschenk CG, Novello V, Pisciotta A, Booyse M, Fouché GW. Integrative effects of vine water relations and grape ripeness level of *Vitis vinifera* L. cv. Shiraz/Richter 99. II. Grape composition and wine quality. South African Journal of Enology and Viticulture. 2014;**35**(2):359-374. DOI: 10.21548/35-2-1022

[70] Bindon K, Varela C, Kennedy J, Holt H, Herderich M. Relationships between harvest time and wine composition in *Vitis vinifera* L. cv. Cabernet sauvignon 1. Grape and wine chemistry. Food Chemistry. 2013;**138**:2-3. DOI: 10.1016/j.foodchem.2012.09.146

[71] Oliveira JB, Egipto R, Laureano O, Castro R, Pereira GE, Ricardo-da-Silva JM. Chemical characteristics of grapes cv. Syrah (*Vitis vinifera* L.) grown in the tropical semiarid region of Brazil (Pernambuco state): Influence of rootstock and harvest season. Journal of the Science of Food and Agriculture. 2019;**99**:5050-5063. DOI: 10.1002/jsfa.9748

[72] Stefanello LO, Schwalbert R, Schwalbert RA, De Conti L, Kulmann MSS, Garlet LP, et al. Nitrogen supply method affects growth, yield and must composition of young grape vines (*Vitis vinifera* L. cv Alicante Bouschet) in southern Brazil. Scientia Horticulturae. 2020;**26**:108910. DOI: 10.1016/j.scienta.2019.108910

[73] Borghezan M, Gavioli O, Pit FA, da Silva AL. Vegetative and productive behavior of grapevines and composition of grapes in São Joaquim, Santa Catarina, Brazil. Pesquisa Agropecuária Brasileira. 2011;**46**(4):398-405. DOI: 10.1590/S0100204X2011000400009

[74] Martínez-Gil AM, Gutiérrez-Gamboa G, Garde-Cerdán T, Pérez-Álvarez EP, Moreno-Simunovic Y. Characterization of phenolic composition in Carignan noir grapes (*Vitis vinifera* L.) from six wine-growing sites in Maule Valley, Chile. Journal of the Science of Food and Agriculture. 2017;**98**:274-282. DOI: 10.1002/jsfa.8468

[75] Anderson K, Nelgen S, Pinilla V, editors. Global Wine Markets, 1860 to 2016 a Statistical Compendium. Adelaide: University of Adelaide Press; 2017. p. 582

[76] Fushing H, Lee O, Heitkamp C, Heymann H, Ebeler SE, Boulton RB, et al. Unraveling the regional specificities of malbec wines from Mendoza, Argentina, and from northern California. Agronomy. 2019;**9**:234. DOI: 10.3390/agronomy9050234

[77] Urvieta R, Jones G, Buscema F, Bottini R, Fontana A. Terroir and vintage discrimination of malbec wines based on phenolic composition across multiple sites in Mendoza. Argentina. Scientific Reports. 2021;**11**:2863. DOI: 10.1038/s41598-021-82306-0

[78] Li H, Li JG, Yang HC. Review of grape and wine industry development in recent 30 years of China's reforming and opening-up. Modern Food Science and Technology. 2009;**25**(4):341-347

[79] Zhan JC, Li DM. The main wine grape varieties. In: Wine Grape Varieties. Vol. 63. Beijing, China: China Agricultural University Press; 2015. 92 p

[80] Sun ZJ. China Wine Year Book 2017. China: Huang Hai Digital Press; 2018. pp. 34-55

[81] Liang N-N, Zhu B-Q, Han S, Wang J-H, Pan Q-H, Reeves M-J, et al. Regional characteristics of anthocyanin and flavonol compounds from grapes of four *Vitis vinifera* varieties in five wine regions of China. Food Research International. 2014;**64**:264-274. DOI: 10.1016/j.foodres.2014.06.048

[82] Cheng G, Fa J-Q, Xi Z-M, Zhen-Wen Zhang Z-W. Research on the quality of the wine grapes in corridor area of China. Journal of Food Science and Technology. 2015;**35**:38-44. DOI: 10.1590/1678-457X.6455

[83] Žurga P, Vahčić N, Banović M, Staver MM. Croatian wines from native grape varieties have higher distinct phenolic (nutraceutic) profiles than wines from non-native varieties with the same geographic origin. 2019;**16**(8):e1900218. DOI: 10.1002/cbdv.201900218

[84] Puškaš V, Jović S, Antov M, Tumbas V. Antioxidative activity of red wine with the increased share of phenolic compounds from solid parts of grape. Chemical Industry & Chemical Engineering Quarterly. 2010;**16**(1):65-71. DOI: 10.2298/CICEQ090714008P

[85] Puškaš V, Miljić U. Effects of fining on phenolic compounds and colour of red wine obtained with addition of increased amounts of grape solid phase in pomace. Hemijska Industrija. 2012;**66**(5):627-634. DOI: 10.2298/HEMIND111209018P

[86] Pantelić M, Dabić Zagorac D, Gašić U, Jović S, Bešlić Z, Todić S, et al. Phenolic profiles of Serbian autochthonous variety Prokupac and monovarietal international wines from the Central Serbia wine region. Natural Product Research. 2018;**32**:2356-2359. DOI: 10.1080/14786419.2017.1408107

[87] Sredojević M. Chemical Characterization of Grape and Monovarietal Wine from Different Vineyard in Serbia. (PhD thesis). Belgrade: University of Belgrade Faculty of Chemistry; 2018

[88] Alvarez-Casas M, Pajaro M, Lores M, Garcia-Jares C. Polyphenolic composition and antioxidant activity of Galician monovarietal wines from native and experimental non-native white grape varieties. International Journal of Food Properties. 2016;**19**(10):2307-2321. DOI: 10.1080/10942912.2015.1126723

[89] Li S-Y, He F, Zhu B-Q, Wang J, Duan CQ. Comparison of phenolic and chromatic characteristics of dry red wines made from native Chinese grape species and *Vitis vinifera*. International Journal of Food Properties. 2017;**20**:2134-2146. DOI: 10.1080/10942912.2016.1233117

[90] De Andrade RHS, Nascimento LS, Pereira GE, Hallwass F, Paim APS. Anthocyanic composition of Brazilian red wines and use of HPLC-UV–vis associated to chemometrics to distinguish wines from different regions. Microchemical Journal. 2013;**110**:256-262. DOI: 10.1016/j.microc.2013.04.003

[91] Granato D, Katayama FCU, De Castro IA. Phenolic composition of south American red wines classified according to their antioxidant activity, retail price and sensory quality. Food Chemistry. 2011;**129**:366-373. DOI: 10.1016/j.foodchem.2011.04.085

[92] Fanzone M, Zamora F, Jofré V, Assof M, Gómez-Cordovés C, Peña-Neira A. Phenolic characterization of red wines from different grape varieties cultivated in Mendoza province

(Argentina). Journal of the Science of Food and Agriculture. 2012;**92**:704-718. DOI: 10.1002/jsfa.4638

[93] Nascimento A, Souza J, Costa A, Santos S, Silva G, Pereira G. Characterization of phenolic composition of altitude tropical wines in the Brazilian northeast. BIO Web of Conferences. 2016;**7**:02014. DOI: 10.1051/bioconf/20160702014

[94] Arcari SG, Caliari V, Souza EL, Godoy HT. Aroma profile and phenolic content of merlot red wines produced in high-altitude regions in Brazil. Química Nova. 2021;**44**:616-624. DOI: 10.21577/0100-4042.20170687

[95] Faustino RS, Sobrattee S, Edel AL, Pierce GN. Comparative analysis of the phenolic content of selected Chilean, Canadian and American merlot red wines. Molecular and Cellular Biochemistry. 2003;**249**:11-19. DOI: 10.1023/A:1024745513314

[96] De Oliveira JB, Egipto R, Laureano O, De Castro R, Pereira GE, Ricardo-da-Silva JM. Chemical and sensorial characterization of tropical Syrah wines produced at different altitudes in northeast of the Brazil. South African Journal of Enology and Viticulture. 2019;**40**(2):1-15. DOI: 10.21548/40-2-3101

[97] King ES, Stoumen M, Buscema F, Hjelmeland AK, Ebeler SE, Heymann H, et al. Regional sensory and chemical characteristics of malbec wines from Mendoza and California. Food Chemistry. 2014;**143**:256-267. DOI: 10.1016/j.foodchem.2013.07.085

[98] Guth H. Identification of character impact odorants of different white wine varieties. Journal of Agricultural and Food Chemistry. 1997;**45**:3022-3026. DOI: 10.1021/jf9608433

[99] Sánchez Palomo E, Díaz-Maroto Hidalgo MC, González-Viñas MÁ, Pérez-Coello MS. Aroma enhancement in wines from different grape varieties using exogenous glycosidases. Food Chemistry. 2005;**92**:627-635. DOI: 10.1016/j.foodchem.2004.08.025

[100] Pérez-Navarro J, Izquierdo-Cañas PM, Mena-Morales A, Chacón-Vozmediano JL, Martínez-Gascueña J, García-Romero E, et al. Comprehensive chemical and sensory assessment of wines made from white grapes of *Vitis vinifera* cultivars Albillo dorado and Montonera del Casar: A comparative study with Airén. Food. 2020;**9**(9):1282. DOI: 10.3390/foods9091282

[101] Elmaci Y, Yıldırım HK, Yücel U, Ova G, Altuğ T. Descriptive profiling of flavor attributes of white wines from different grape varieties. International Journal of Food Properties. 2007;**10**:651-659. DOI: 10.1080/10942910601098072

[102] Garrido-Bañuelos G, Ballester J, Buica A, Mihnea M. Exploring the typicality, sensory space, and chemical composition of Swedish solaris wines. Food. 2020;**9**:1107. DOI: 10.3390/foods9081107

[103] Winestats. Compositional changes to the vineyard in England & Wales. 2014. [Internet]. Available from: http://www.winestats.ingo/composition-vineyards-england/ [Accessed: May 10, 2022]

[104] Boncinelli F, Contini LCC, Gerini F, Scozzafava G. The consumer loves typicality but prefers the international wine. Agriculture and Agricultural Science Procedia. 2016;**8**:236-242. DOI: 10.1016/j.aaspro.2016.02.098

Chapter 3

Overview of the Esca Complex as an Increasing Threat in Vineyards Worldwide: Climate Change, Control Approaches and Impact on Grape and Wine Quality

Evangelos Beris, Moustafa Selim, Despoina Kechagia and Alexandra Evangelou

Abstract

Esca is an increasing threat to global viticulture causing significant losses by reducing yields, declining or wilting vines, and shortening the productive life of vineyards. Recent findings indicate that the disease may also affect the quality of grapes and the chemical composition of musts and wines. However, more research in this field is needed. Esca seems to affect the ripening process of grapes resulting in lower sugar content, higher acidity, and increased nitrogen concentrations. Regarding polyphenolic compounds, reduction on the concentrations of (+)-catechin, (−)-epicatechin, anthocyanins, and tannins has been observed due to the alteration of flavonoid metabolism. Esca is a complex-chronic disease, where several fungal pathogens act simultaneously or successively, to cause necrosis to the vascular tissues of grapevines by blocking the xylem vessels and by producing enzymes and phytotoxic metabolites. As genotype affects stress response, specific *Vitis vinifera* cultivars present higher levels of resistance to the disease than others. There is evidence that varieties such as Merlot, Grenache Rouge, and Roussanne are relatively resistant, compared with more susceptible Cabernet Sauvignon, Mourvèdre, Sauvignon Blanc, and Semillon. Another main objective of the current work was to investigate the possible effects of climate change on Esca development and propose appropriate control strategies.

Keywords: Esca, *Vitis vinifera*, plant stress, fungal pathogens, grapevine, wine quality, must composition, climate change

1. Introduction

Grapevine trunk diseases (GTDs) are an increasing threat to global viticulture [1, 2]. The causal agents of GTDs belong to diverse groups of both ascomycetes and basidiomycetes that colonise mainly the vascular tissues of grapevines, interfering with plant physiology, microbial ecology and activating plant response mechanisms [3, 4]. The Esca complex is considered as one of the most common and catastrophic wood diseases of the European grapevine "*Vitis vinifera* L." [5]. Losses and damages caused by the disease are currently in an increasing rate, especially in warm viticultural regions. Therefore, it might be assumed that climate change influences the development of Esca and the responses of the host towards the pathogens [6]. Mature vines (over 15 years old) present greater possibility to be affected by Esca and other GTDs [7]. It is generally known that old vines are responsible for the production of some of the world's finest wines, expressing high quality and typicity according to their region [8]. Therefore, Esca is considered as a serious threat, especially in regions where the production of wine from old—usually ungrafted—vines is economically important. Other forms of the disease occur in young grapevines and are usually associated with contaminated propagating material [9, 10]. Esca is considered as a complex disease, as it is caused by the action of several fungal pathogens infecting the vines in various manners. However, it has been proposed that Esca can also be described as a complex of different diseases that coexist on the trunk of mature grapevines [11, 12]. Some issues regarding aetiology, epidemiology, the role of pathogens and the exact factors that influence the development of the disease remain unclear.

Typical symptoms of Esca include inner wood necrosis, either as brown-black streaks caused by ascomycetes (although teleomorphs are not known for all these fungi), or as white rot (a yellowish-white discoloration of the wood, which is soft, spongy and brittle) caused by basidiomycetes, while in the leaves, peripheral and interveinal chlorosis, which results in necrosis (tiger stripes), is often observed in the summer due to insufficient water supply and the presence of phytotoxic metabolites [13, 14]. However, the latter symptom is often described as a separate disease within the Esca complex and is called "grapevine leaf stripe disease" (GLSD). In severe cases, gradual necrosis of arms may be observed or even sudden wilt and death of the entire vine (apoplexy). Esca is found in most viticultural regions worldwide, and the disease is believed to occur for as long as grape growing exists [11].

Climatic and soil parameters significantly affect the frequency of Esca-related symptoms and implications [15–17]. Drought and temperature are the most crucial abiotic factors for the enhancement of the pathogens affecting growth rates, propagule germination and the rates of inoculum production [18]. Intense heat and drought increase the respiration rates of plants, causing carbon losses and even plant death [19, 20]. In general, water availability induces modifications of vessel diameter in various plant species [21, 22]. The worsening of the symptoms observed in Esca-infected plants may be partially due to the size of the vessels that have been formed by altering water supply [23]. Research on the possible pathological nature of Esca began in France in 1898 [24]. However, great progress and additional findings regarding aetiology, pathogenesis and diagnosis have been revealed only recently. Abiotic factors such as climate, soil type and topography as well as biotic factors such as age of the vines and cultivar (even clone) appear to affect disease incidence [2, 7]. Further research is needed to thoroughly investigate all the aspects of aetiology, ecology and conditions that favour the development of the disease, as this knowledge can contribute to designing control strategies, while it is known that conventional chemical control is not easily applicable in the case of Esca [17, 25, 26]. Moreover, the impact of climate change needs to be addressed, as Esca is an increasing threat in vineyards worldwide and water stress seems to play a role [27].

2. Pathogens - Syndromes

A complex of several fungi coexists to variable levels in inner vessels, causing certain syndromes. It is commonly agreed that two species of ascomycetes, *Phaeoacremonium minimum* (previously known as *Phaeoacremonium aleophilum*) and *Phaeomoniella chlamydospora* as well as one species of basidiomycetes, *Fomitiporia mediterranea*, are the predominant fungi involved in the development of Esca on a global scale [28, 29]. *P. minimum* and *Pl. chlamydospora* are responsible for the brown streak syndrome of the wood, while wood decay (white rot) is caused by *F. mediterranea* and other basidiomycetes. Nevertheless, it has been found that many more fungal species are involved in the Esca complex or in Esca-related symptoms and damages. In addition to *F. mediterranea*, other Basidiomycetes have been reported by Fischer and González-García [30] either as primary pathogens (*Inonotus hispidus*) or as secondary infections developing on already dead tissues of grapevine trunks, causing white rot (*Stereum hirsutum*, *Trametes hirsuta* and *Schizophyllum commune*). *F. mediterranea* is a newly found species [31], identified after molecular analysis was conducted to isolates that were previously believed to belong to *F. punctata*. Those isolates differed from *F. punctata*, and therefore, a new species *F. mediterranea* (M. Ficher) was established and believed to exist widely in grapevines across the Mediterranean basin [29, 32]. Different *Fomitiporia* species have been associated with Esca in north America (*F. polymorpha*), Australia and New Zealand (*F. australiensis*), while other basiodiomycetes (*Inocutis jamaicensis* and *Fomitiporella vitis*) have been found to cause white rot in South America [33]. In South Africa, a diverse group of basidiomycetes was found in grapevine trunks, including fungi of the genera *Inocutis*, *Inonotus*, *Phellinus* and *Fomitiporia* [34, 35].

Regarding ascomycetes, a diversity of fungi, mainly within the order Diaporthales, exists in grapevine trunks. However, *P. minimum* and *Pl. chlamydospora* are considered as the most common species in both Esca and Petri disease. Many *Phaeoacremoniun* species have been characterised as vascular plant pathogens, producing enzymes and phytotoxic metabolites, causing dieback and wilt to several woody hosts [36, 37]. Multiple genotypes of *P. minimum* can be found within the micro-environment of a single vineyard [38]. At least 24 species of the genus *Phaeoacremonium* have been reported to infect grapevines in a global scale [39]. In South Africa, there is extended diversity of *Phaeoacremonium* species probably due to the large flora diversity, and newly found species, include *P. album*, *P. aureum*, *P. bibendum*, *P. gamsii*, *P. geminum*, *P. junior*, *P. longicollarum*, *P. meliae*, *P. oleae*, *P. paululum*, *P. proliferatum*, *P. rosicola* and *P. spadicum* [37]. Most of those species have been found in grapevines and to lesser extend in other tree crops.

According to a classic definition [12], diseases within the Esca complex are distinguished into five typical syndromes:

a. *Brown wood-streaking*: This form causes damages to rooted cuttings and rootstocks, usually in nurseries. It is associated with Ascomycetes of the genus *Phaeoacremonium* (or related genera), often without showing external symptoms.

b. *Petri disease*: It is caused by the early infection of propagating material or young grapevines by species of *Phaeoacremonium* or related genera.

c. *Young vine Esca*: Fungi of the genus *Phaeoacremonium* infect young grapevines (up to 8–10 year old) through wounds, and they cause black or brown wood-streaking and vascular gummosis inside the trunk, with or without foliar symptoms.

d. *White rot (Esca)*: When infection through wounds is caused exclusively or mainly by Basidiomycetes (such as *F. punctata*, *F. mediterranea*, etc.), a spongy wood decay (white rot) occurs, which may or may not be associated with leaf and fruit symptoms.

e. *Esca proper*: White rot develops in the trunk of old vines, simultaneously or after the development of brown wood-streaking. This full-scale syndrome is caused by the combined or sequential action of both ascomycetes and basidiomycetes.

3. Geographical distribution of Esca

Grapevine trunk diseases, especially Esca, are increasing in incidence and severity within vineyards and nurseries worldwide [3] and threaten the wine industry globally [40]. It has become increasingly devastating during the past three decades and represents today a major concern in all wine-producing countries [41]. Over the past two decades, GTDs (especially the Esca complex) have been reported in European, American and South African vineyards [35, 42]. A survey on GTDs was carried out during 2015 and 2016 in 18 European and four Mediterranean countries on a regional and a national level. Result showed that both chronic form and apoplexy occurred in all the surveyed countries [42]. However, not all the viticultural regions are affected equally. A dramatic increase of the disease has been reported in some Mediterranean regions such as Tuscany, where more than 50% of the vineyards have a disease incidence ranging from 20% to 30% [43]. Similar observations have been made for France, Greece or Portugal [11, 44–47]. The estimated annual increase is estimated to be 4–5% [11].

In many regions, implications caused by Esca have a substantial impact in viticulture and wine production. It has been recently observed that Esca affects numerous grapevines in the volcanic island of Santorini (Cyclades—Greece) and many more Greek islands including Crete. Santorini is considered as the most important white wine region in Greece, producing distinctive and terroir driven wines from old ungrafted vines (up to 250 year old) on a phylloxera safe volcanic soil. Increasing losses and damages have been recently reported in many regions across the Mediterranean basin. Nevertheless, current research indicates that Esca is an increasing phenomenon on a global scale, especially in wine regions that depend on the production of wines from mature vines. Esca is also gaining increasing importance in Central European wine-growing countries [32].

4. Infection - Epidemiology

Infections are mainly caused by fungal spores produced on infected or dead vines and transmitted via wind or rainfall. Spores usually invade their host through wounds created by various causes such as pruning, severe frosts and mechanical damages to the trunk or root system [48]. However, pruning incisions are considered as the main portals of infection [15, 49]. It has been recently recorded that arthropods, such as millipedes, ants, spiders or even beneficial predatory insects, can vector pathogens related to Esca and other GTDs, via wounds of freshly pruned grapevines [50]. Regarding the forms of the disease that occur in young grapevines, the infection is usually associated with the use of contaminated propagating material or insufficient hygiene conditions and practice in nurseries [9, 10].

P. minimum and *Pl. chlamydospora* are characterised by their aerial dispersion [51, 52]. The release of *Pl. chlamydospora* spores is associated with rainfall, while *P. minimum* occurs during the bud brake period without the presence of rainfall [53, 54]. Sources of infection have been observed in protected areas of wood within deep cracks [55, 56]. Both *Pl. chlamydospora* and *P. minimum* can also invade their hosts via grafting or spread through contaminated plant material [9, 57]. These infections are associated with Petri disease, but it is possible that the pathogens remain in a latent form and appear later for the development of Esca proper. As *F. mediterranea* and *P. minimum* reproduce sexually, basidiocarps and perithecia, respectively, may represent sources of inoculum in the field [38, 43, 56]. Many studies indicate the existence of pathogens that do not directly manifest the disease (latent infection). Factors that affect pathogenesis on *V. vinifera*, as well as the appearance of symptoms, include age of vines, training system, size of the pruning wounds, time of pruning, climatic conditions and susceptibility of certain varieties.

Esca is considered as a catastrophic, non-selective grapevine disease, and remedies such as specialised fungicides or totally resistant varieties do not exist [17, 58]. Nevertheless, various *V. vinifera* cultivars present different levels of susceptibility or resistance to the pathogens. In a study conducted in Greece, 'Agiorgitiko' and 'Soultanina' displayed susceptibility, while 'Assyrtiko' and 'Xinomavro' showed resistance to *Pl. chlamydospora* [59]. However, in homologous experiments that took place in Spain, all the varieties (including 'Godello', 'Albarín Blanco', 'Doña Blanca', 'Pan y Carne', 'Palomino', 'Zamarrica', 'Mencía' and 'Brancellao') were susceptible to the fungus [60]. Moreover, the major Spanish red grape Tempranillo is known to be very susceptible to most types of Esca [17]. In previous bioassays conducted near Siena, Italy, several *V. vinifera* varieties were tested in terms of symptom severity. Results indicated that the 17 vine cultivars were categorised into four groups of varying susceptibility to Esca. Semillon was shown to be the most susceptible cultivar, while Roussanne was the most resistant [61]. The above study also suggested that various rootstocks can affect susceptibility to Esca. In other experiments, Esca caused reduction (about 70%) in carbon assimilation and stomatal conductance. These effects were greater in varieties such as Cabernet Sauvignon and Sangiovese compared to Trebbiano [58]. Regarding table grape varieties, a previous study indicated that *V. vinifera* cv. Matilde was more resistant compared to *V. vinifera* cv. Italia, in terms of internal symptoms' severity [62]. Thompson seedless was also evaluated as relatively susceptible [63]. Even at a clonal level, there might be a difference on the expression of the disease. A study in France confirmed a clone-dependent expression of Esca in Chardonnay [2]. Overall, important international wine-grape varieties such as Cabernet Sauvignon, Sauvignon blanc, Trebbiano, Mourvèdre and Cinsault are more susceptible to Esca compared to the relatively resistant Pinot Noir, Merlot, Grenache, Carignan and Roussanne [55, 63–65].

5. The effects of climate change on Esca development

Pathogens may be established in grapevines at early stages but remain in a latent form. The impact of the disease becomes greater as the vines mature. Older plants also experience more infection cycles via pruning wounds [7, 66, 67]. Since the time of trunk infection, and until the onset of symptoms, the disease progresses slowly. As latent pathogens, fungi involved in the development of Esca begin their life cycle inside plant vessels but without causing any symptoms. Years post infection, fungi become more invasive resulting in symptom appearance [68].

Abiotic factors, such as environmental conditions, are also recorded to significantly influence the development of the disease and the time of symptom manifestation [15, 16, 58]. Spore infections are favoured by intense humidity. Since most Esca-related fungi are anaerobic, intense oxygenation inhibits their growth, while carbon dioxide favours it. The optimal growth temperature for most of those fungi is 20–25°C. Research regarding the effects of abiotic factors on Esca was conducted at three commercial Tempranillo vineyards located in three different wine regions of Spain [17]. Those viticultural regions differ with each other in terms of average annual temperature, lowest and highest temperatures during the coldest and warmest months, respectively, climate classification and planting arrangements. The area that presented the highest vine infection (30%) is characterised by relatively high annual temperature and quite high temperatures during the hottest months, but normal lower temperatures in the coldest months. The climate of the region was described as warm, extremely humid with hot summers. The planting distance was 2 × 3 m (low planting density).

Symptomatic vines often present significantly lower leaf photosynthesis rates [69, 70]. A progressive decline in net photosynthesis was observed in asymptomatic leaves from symptomatic shoots [71]. Reduced photosynthesis at the early stages of GLSD symptoms might be a result of stomatal closure and can be magnified under water stress and high-temperature conditions. Alterations in carbohydrate physiology of grapevines may begin with various stress types such as low temperatures, drought or chemicals and can lead to considerable disorders of carbon sustenance, as a result to Esca infection. Studies indicate that highly infected grapevines may perceive some signals and react precociously by reducing photosynthesis and by triggering various defence mechanisms [71, 72].

Most Esca symptoms appear on vines during summer, where transpiration rates and demand for water increase [48, 73]. It is known that when plants are exposed to various stresses, the expression of several genes is exhorted. Consecutively, enzymes and hormones with multiple biological functions are produced due to environmental stimuli. Therefore, toxicity from the action of pathogens, host defence responses, and environmental conditions compose the disease triangle [6] (**Figure 1**). Exceptionally hot and dry weather periods have been repeatedly recorded worldwide during the past two decades. Among these, the most remarkable in Europe have been the summers of 2003, 2006, 2011 and 2018. Most infected plants remain symptomless externally for years after infection. Appearance of external symptoms is not directly related to inner wood decay and may be associated with cultivation practices or environmental stresses such as high temperature, water shortage or overstock. Water stress has been proven to significantly influence characteristics such as shoot length, stem weight, expression of wood symptoms and plant survival, when young vines were artificially inoculated with conidial suspensions of *Pl. chlamydospora* [74]. Moreover, in the case of *V. vinifera*, water stress has been recorded to eventually alter the size of vessels and to affect xylem hydraulic conductivity, resulting in reduced water flow rates [22, 23, 75]. This is crucial for the outbreak of the symptoms. Foliar symptoms may also be linked to stress-related pathways of *V. vinifera*, which are stimulated during water, salt and oxidative stresses [76]. However, the effects of environmental stresses on the transcriptional responses of grapevines to the Esca complex have not been fully explored yet. Researchers

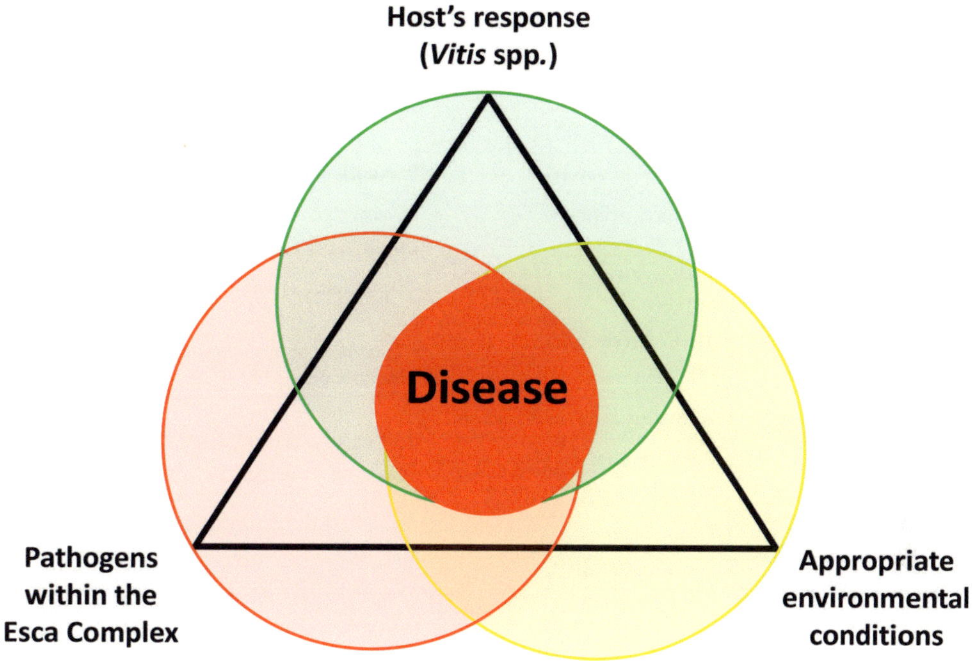

Figure 1.
Depiction of the disease triangle (combined factors that cause the disease) in the case of Esca.

investigated the role of water stress-related genes while studying the early events that take place before apoplexy [72]. None of the tested genes was affected in visually healthy pre-apoplectic leaves, but they were down-regulated in drying leaves. Distribution and symptomatology of most GTDs are associated with the exposure of vines to different climatic conditions [15]. The influence of heat and water stress on grapevine physiology was recorded in the case of Eutypa dieback [77]. Back to the case of Esca, correlation between rainfall, high temperatures and the severity of GLSD symptoms was observed in recent studies conducted in Italy [16].

6. Detection methods

It is important to use detection methods to detect the presence of Esca pathogens before the onset of symptoms (**Table 1**). This will enable grapevine nurseries to stop or at least limit the spread of infection. Using detection methods, it was reported that infection may take place in nurseries during the propagation process [78], storage [79] or due to the use of infected mother plants [80, 81], while rootstocks used for propagation have been reported to harbour Esca pathogens [82, 83]. Tools used in nurseries such as hydration tanks, grafting tools and callus media [84–86] were also shown to spread infection. Therefore, detection methods will help to improve the sanitary measurements to avoid contamination in these facilities. Molecular detection tools can also provide markers for Esca pathogens. These molecular markers can be used as tools in taxonomy allowing the discrimination on strain level [87]. Examples for such tools are sequencing of the internal transcribed spacer region (ITS) [29, 31, 88], restriction fragment length polymorphism (RFLP) [43, 88] and sequence-characterised amplified region (SCAR) [89]. Molecular markers shed light on the genetic variation among

Detection and control of Esca		
Detection methods	**Molecular**	**Physical**
	Molecular markers: ITS, RFLP, SCAR, RAPD	Spectral: hyperspectral, multispectral
	Prevention in nurseries	**Prevention in vineyards**
Control methods	• Chemical treatment of cuttings • Hot water treatment of cuttings • Antagonist treatment of cuttings	• Use of clean planting material • Removal of infected parts • Optimizing pruning method (method and period) • Treatment of pruning wound chemical or biologically
	Curative measures in vineyards	
	• Trunk renewal • Over-grafting • Trunk surgery • Trunk injection	

Table 1.
Synopsis of detection methods and control strategies to prevent/lessen Esca.

Esca pathogens over geographical areas as well as way of dispersion. Genetic variation among isolates of pathogens was investigated with random amplified polymorphic DNA (RAPD) [89, 90]. It was suggested that variation within species may be related to the geographic location of the isolates [90, 91]. It has also been suggested that *F. mediterranea* is spread via airborne basidiospores and that outcrossing occurs [90]. Further studies on the occurrence of *Pl. chlamydospora* in different regions as well as the use of additional markers may provide more information on introduction frequencies, geographical spread and to determine the mechanisms of inoculum dispersal [38, 92]. Another molecular tool in Esca research is the whole genome sequencing. It is considered one of the most important tools in understanding the biology of organisms. Recently, the genome sequence of the two most common and important Esca pathogens, *Pl. chlamydospora* and *P. minimum*, has been released [93, 94]. These results will reveal the genetic makeup of these two pathogens, which, in turn, will help discovering new genes involved in pathogenicity as well as markers that could further differentiate between species.

Another detection method is spectral imaging using sensors to detect foliar symptoms before being visible to the human eye using ground-based hyperspectral and airborne multispectral imaging. Since chronic symptoms do not develop every year [3], an annual monitoring is a fundamental tool to estimate disease incidence in the vineyard [95]. Leaf measurements can either be recorded in the whole spectral region (hyperspectral) or at selected spectral bands only (multispectral) [96]. According to a recent study, external symptoms could be detected pre-symptomatically; however, the authors emphasised that these results need further evaluation [95]. A similar imaging approach using RGB images was used to detect Esca symptoms. The detection accuracies were 88% and 91% for white and red cultivars, respectively [97]. The reason why spectral imaging seems to be promising is the fact that net photosynthesis rate was shown to decrease gradually in asymptomatic leaves [69, 71]. However, it is not known so far how early can spectral imaging techniques detect Esca infection.

7. Conventional and modern control approaches

Control strategies against Esca mainly include preventive measures and cultivation practices (**Table 1**). In many wine regions across the Mediterranean Sea, sodium arsenite was extensively used as a chemical control mean. However, sodium asrenite is now banned in most countries. Other compounds, such as copper oxychloride, benodanil, fosetyl-Al, hydrogen peroxide, glutaraldehyde and furmetamide, have also been tested as inhibitors of Esca. Nevertheless, it is certain that wood decay is irreversible, and there are no remedies or conventional control methods to cure Esca [17, 25]. The use of a mixture consisting of calcium chloride, magnesium nitrate and Fucales seaweed extract resulted in a significant reduction of symptoms in Trebbiano d'Abruzzo and Montepulciano d'Abruzzo vineyards [70]. This application also seemed to increase the quantity and quality of grapes, without causing any phytotoxic effects on the berries. It is currently believed that research should focus on the use of incorporating solutions that could activate plants' natural defence responses, as well as studies of molecules that exhibit antimicrobial properties. The use of relatively resistant cultivars and healthy propagating material are considered as preventing means of great importance [48].

Preventive measures in nurseries: The use of fungicides to control Esca pathogens in nurseries is difficult. Chemical dips are usually used to control external pathogens, but they do not penetrate grapevine cuttings and, hence, are ineffective against pathogens inhabiting the vascular tissues [98]. Indeed, Chinosol (hydroxyquinoline sulfate) was reported to be the most commonly used fungicide; however, it was reported to be ineffective against *Pl. chlamydospora* and *P. minimum* in grapevine nurseries [99]. Treating propagation material with hot water is considered the most effective method to disinfect dormant canes during the propagation process [98, 100]. However, some reports showed negative side effects on grapevine growing from these cuttings such as poor shoot development and less rootstock vigour [101], delayed callusing and rooting of cuttings [98], delayed development or bud death in cuttings and grafted vines [99, 102, 103], and incomplete healing of graft unions and fermentation in cold storage [104]. This, in turn, leads other authors to suggest that cuttings taken from vines grown in warm climates may tolerate hot water treatment (HWT) better than cuttings taken from vines grown in cool climates [105]. HWT can be applied to rootstock cuttings prior to grafting [73, 106] or to young, grafted vines just prior to dispatch [106, 107]. Another method is the use of biological control agents such as *Trichoderma* in various formulations such as powder, granules/pellets and dowels [108]. Incidence of *Pl. chlamydospora* and *Phaeacremonium* spp. in rootstock cuttings was reduced by soaking the planting material in *Trichoderma* formulations [106]. Moreover, it was demonstrated that the application of *Trichoderma atroviride* at hydration, callusing and pre-planting stages in nurseries reduced infection by *Pl. chlamydospora* and *P. minimum*, hydration treatments being the most effective [109].

Preventive measures in vineyards: The first line of preventive measures is the use of pathogen free material obtained from healthy vines. However, there is no guarantee since asymptomatic vines may harbour Esca pathogens. For this reason, many nurseries treat cuttings with fungicides, hot water or biological agents to ensure that cuttings are pathogen free. Another preventive measure is the phytosanitation, which includes the destruction of affected parts, pruning debris or whole vines. Infected pruning residues should be immediately removed from the vineyard and destroyed since fungal fruiting bodies can become a potential source of inoculum for new infections [110]. Since pruning wounds are the major entry point for Esca pathogens [108, 111], keeping the wounds at minimum and treating these wounds are very important for protecting the vines before spores germinate and start colonising the open xylem vessels and pith parenchyma cells at wound area [50, 112]. Another important aspect is to avoid pruning during rainy or humid warm days as spores are released under these environmental conditions [113, 114]. Late pruning (prior to budbreak) is also recommended because wound healing is rapid during this time [115]. Moreover, double pruning [116], which consists of mechanical pruning in the winter followed by hand pruning before budbreak, is a viable technique to complete pruning in large vineyard areas [117]. Wound treatment can be done chemically through chemical sealant or biologically by an antagonist. Several compounds to treat pruning wounds were tested in paste forms such as thiophanate-methyl (Topsin M®), cyproconazole + iodocarb treatment (Garrison®—commercial tree wound past), boric acid in a wound sealant paste (Biopaste®) and a pyraclostrobin formulation (Cabrio®) and showed positive results as wound protectants [73]. Others were tested as sprays such as myclobutanyl that gave positive results against Esca pathogens [118]. Protecting the wound biologically can be done by *Trichoderma* sprayed on the fresh pruning wounds one to two days after pruning. Results showed a 45–65% reduction of leaf symptoms in treated vines when compared to the untreated vines [119]. Last but not least, balanced nutrition and irrigation are also important, as preventive measures, as they enable vines to fight the infection [6].

Curative measures in vineyards: There are several ways to manage infected vines; however, some of these measures cannot guarantee the removal of Esca pathogens completely. These methods include trunk renewal, over-grafting, trunk surgery and trunk injection. Trunk surgery is an old practice especially in fruit trees in the Mediterranean area. Surgery is a technique that relies on cutting the trunk open with a chainsaw and removing of decayed wood from the heart of the trunk allowing quick recovery of symptomatic vines by keeping their active root systems and, therefore, maintaining the quality of the product that is linked to the vine age [120]. Trunk surgery is considered more expensive and time consuming compared to trunk renewal [26] and needs well-trained personnel [120] not to compromise the main sap flow during surgery, preferring to act only on one side of the plant [121]. Another curative measure is to cut back the trunk below infected wood until healthy/clean wood is visible. One of the healthy suckers (water shoots) is then trained up to become a replacement trunk. These suckers arise from base buds at prior node positions on the vine trunk. This technique also works with other trunk diseases. It presents the advantage of saving the vine root system. Training suckers to new trunks and arms may precede trunk removal; therefore, no yield needs to be lost [122]. If suckers are not available, an over-grafting or re-grafting may be done, while after cutting the vine and reaching healthy tissues, the vine is then over-grafted or re-grafted. According to some empirical experimentation, the treated vines could reach productivity in 3 years, because of their mature root system [123].

Curative chemical treatment seems difficult since Esca pathogens exist in the trunk. However, in recent years, some methods have been developed to deliver fungicides in the trunk using trunk injection, also called endotherapy. Early trunk injections on Esca-infected vines showed negative results [25, 124]. However, these studies used exclusively the presence of leaf symptoms to evaluate treatment success and did not examine changes in the presence of pathogens in the wood pre- or post-treatment [125]. A new injection technique based on encapsulated nanoparticles has shown promising results. Several fungicides have been already successfully encapsulated for being delivered in the trunk. In a recent study, fungicides were coated with lignin to attract Esca pathogens. Indeed, lignin-coated fungicides successfully inhibited the growth of *Pl. chlamydospora* and *P. minimum* even after a period of four years, proving a drastic reduction in Esca leaf symptoms after a single injection [126]. Trunk injection encouraged some winegrowers to inject many compounds such as hydrogen peroxide into grapevine trunks; however, more testing is required before this technique can be recommended.

8. Impact of Esca on the quality of grapes and wines

Grapevines infected by Esca-related pathogens do not consistently demonstrate foliar and fruit symptoms every growing season [3, 61]. However, those vines tend to present decreased vigour and lower yields. The Esca complex disease causes significant losses in wine production worldwide [58]. Esca has also been reported to affect grapes' quality and, subsequently, the quality of wines produced from those grapes [108] (**Figure 2**). In many cases, quality degradation can even be observed on grapes macroscopically. As a result of chronic development of Esca, small dark-brown-to-purple spots, described as 'black measles', often appear on the berries [36]. The ripening process of grapes is also affected indirectly, as Esca-infected vines present altered rates of photosynthesis [69, 71].

In a study conducted in Bordeaux, France, levels of ripeness were significantly delayed in vines infected by Esca. In addition to reduced yield, a decline of the sugar content (~10%) as well as higher acidity (~20%) was observed. The organoleptic characteristics were also affected, as wines partially made from Esca-infected vines

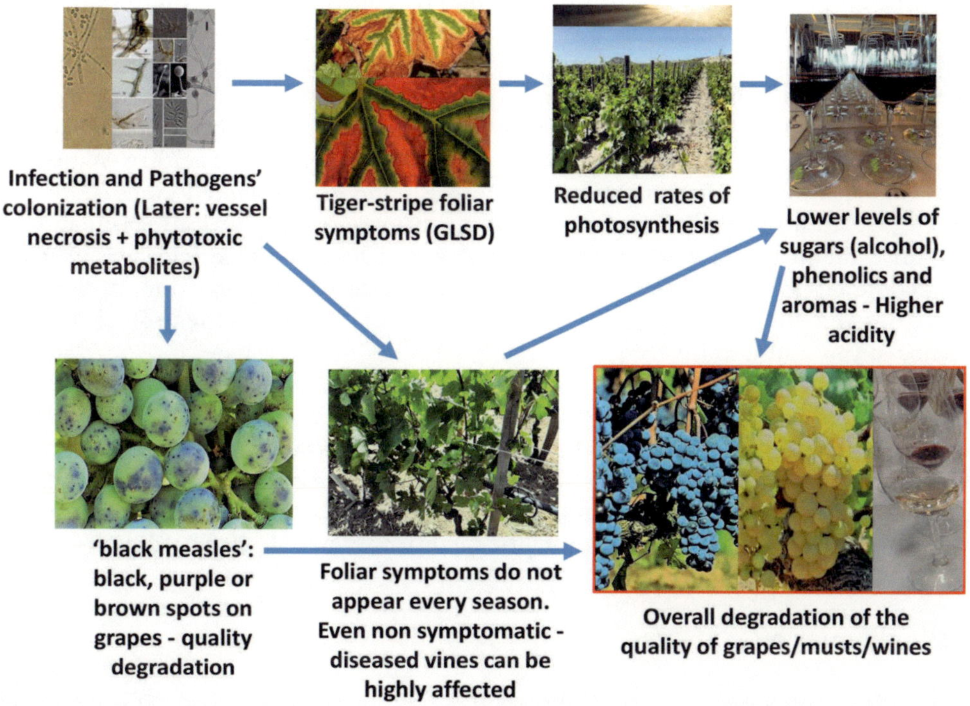

Figure 2.
Development of Esca and its implications on the quality of grapes and wines.

presented lower levels of fruity aromas and elevated levels of vegetal-herbaceous characters [127]. In another research that took place in Italy, the effects of Esca on the quality of grapes and the composition of wines were tested in two Trebbiano d'Abruzzo vineyards [128]. Three different groups were evaluated: (a) grapes from diseases and symptomatic vines, (b) grapes from vines that were asymptomatic at the time of experiments but were previously confirmed to be diseased and (c) healthy grapes from healthy vines. The results indicated notably lower levels of sugars and substantially higher nitrogen concentration in the berries from diseased-symptomatic grapevines compared to the other two groups (which did not differ significantly with each other). Regarding polyphenolic concentrations, significant differences between healthy and unhealthy vines were also observed, but results were conflicted between the two vineyards. Consequently, wines produced from diseased-symptomatic vines displayed lower alcohol content by ~1% as well as higher levels of tartaric and malic acid and higher total nitrogen. However, no significant differences were detected in the cases of pH and total acidity [128]. Subsequent experiments were performed in the same vineyard areas (Controguerra and Giulianova in Teramo, Italy). In the latter bioassays, notably higher levels of trans-resveratrol as well as iron and magnesium levels in musts from diseased-symptomatic vines were observed [25]. Symptomatic vines also produced higher concentrations of nitrogen confirming the results of 2001. This constant increase of nitrogen may be a result of protein hydrolysis that occurs in diseased leaves and generates amino acids that migrate to the grapes. Moreover, grapes from symptomatic vines presented significantly lower sugar levels and musts from symptomatic vines had significantly higher levels of total acidity, malic acid and total polyphenols compared to the musts from the other groups.

High-performance liquid chromatography was used to detect reduction in concentrations of (+)-catechin, (−)-epicatechin and anthocyanins on skins of Cabernet Sauvignon grapes, from Esca-infected vines, in the region of Bordeaux during 2009 and 2010 [65]. However, no statistical difference was detected in the case of total tannins. In wines, significantly lower alcohol content was produced from infected vines, but there was no difference in terms of total acidity. Wines were made using various percentages of berries from Esca-infected vines (0, 5, 15, 25, 50, 75 and 100%). The phenolic composition of grapes was affected, and the sensorial quality of wines was decreased, when Esca-infected fruit was used in percentages as low as 5%. It was reported that Esca affects normal maturation of berries and alters the flavonoid metabolism, which regulates the production of anthocyanins and tannins [65].

In the case of Sauvignon Blanc, symptomatic vines produced grapes of inferior quality, but when curettage was applied, it seemed to have a positive effect. Curetted and asymptomatic grapevines produced wines with favourable chemical and organoleptic characteristics, maintaining high quality and typicity of Sauvignon Blanc [5]. Overall, Esca seems to reduce sugars, aromas and some phenolic compounds, resulting in the deterioration of the organoleptic quality of wines [4, 25, 65, 128].

9. Final remarks

Esca is a complex disease. Several pathogens and factors act in association or in succession, to cause the syndromes, which become fully expressed only under specific environmental conditions. The appearance of symptoms as well as the severity of the disease seems to be substantially affected by abiotic factors. Temperature increases transpiration and, therefore, is a key factor, affecting not only the time of symptom appearance but also infection, pathogenesis, fungal growth and the level of damage. Esca is currently considered as an increasing threat, becoming severe in many viticultural regions across the globe. Climate change might be a reason for the dispersal and harshness of the disease, as losses and damages are associated with plant responses to various stresses.

Control approaches include preventive measures in both vineyard and nurseries, as well as curative techniques in Vineyards. Despite the progress, direct control remains difficult. Certain varieties present some levels of resistance to Esca, compared to others, but there are no varieties with immunity to the disease. The reasons behind susceptibility or resistance are probably related to some interactions between Esca pathogens and grapevine physiology, but this phenomenon is still poorly understood. Most publications indicate resistance of specific varieties to a single pathogen only, without examining the response of various cultivars to the entire complex under field conditions.

It is known that Esca causes substantial economic losses to global viticulture by reducing yields and shortening the productive life of grapevines. Moreover, it seems to affect the quality of grapes and wines as infected vines (both symptomatic and asymptomatic) tend to produce less concentrated grapes with lower amounts of sugars and phenolic compounds. The fact that even asymptomatic vines produce grapes of inferior quality must be seriously taken into consideration. However, further research is needed to examine accurately and precisely the effects of Esca on the quality of grapes and the chemical composition of musts/wines. Overall, additional research must be conducted to investigate all the aspects of vine physiology and plant responses, in regard to ecology, pathogenesis, lifecycle, physiology and diversity of the fungal pathogens that are involved in the Esca complex.

Acknowledgements

Sincere gratitude to the German Academic Exchange Organisation (DAAD) for supporting the memorandum of co-operation between the department of Wine, Vine & Beverage Sciences of the University of West Attica, Greece, and the University of Geisenheim (HGU), Germany.

Author details

Evangelos Beris[1]*, Moustafa Selim[2], Despoina Kechagia[1]
and Alexandra Evangelou[1]

1 Department of Wine, Vine and Beverage Sciences, School of Food Sciences
University of West Attica, Aigaleo, Athens, Greece

2 Department of Crop Protection, Hochschule Geisenheim University, Geisenheim, Germany

*Address all correspondence to: eberis@uniwa.gr; evansagri@hotmail.com

References

[1] Gramaje D, Úrbez-Torres JR, Sosnowski MR. Managing grapevine trunk diseases with respect to etiology and epidemiology: Current strategies and future prospects. Plant Disease. 2018;**102**:12-39. DOI: 10.1094/PDIS-04-17-0512-FE

[2] Moret F, Lemaître-Guillier C, Grosjean C, Clément G, Coelho C, NegrelJ, et al. Clone-dependentexpression of Esca disease revealed by leaf metabolite analysis. Frontiers in Plant Science. 2019;**9**:1960. DOI: 10.3389/fpls.2018.01960

[3] Bertsch C, Ramírez-Suero M, Magnin-Robert M, Larignon P, Chong J, Abou-Mansour E, et al. Grapevine trunk diseases: Complex and still poorly understood. Plant Pathology. 2013;**62**:243-265. DOI: 10.1111/j.1365-3059.2012.02674.x

[4] Bruez E, Vallance J, Gerbore J, Lecomte P, Da Costa JP, Guerin-Dubrana L, et al. Analyses of the temporal dynamics of fungal communities colonizing the healthy wood tissues of esca leaf-symptomatic and asymptomatic vines. PLoS One. 2014;**9**:95928. DOI: 10.1371/journal.pone.0095928

[5] Bruez E, Cholet C, Thibon C, Redon P, Lacampagne S, Martignon T, et al. Influence of curettage on Esca-diseased *Vitis vinifera* L. cv. Sauvignon blanc plants on the quality of musts and wines. OENO One. 2021;**55**(1):171-182. DOI: 10.20870/oeno-one.2021.55.1.4479

[6] Fischer M, Peighami-Ashnaei S. Grapevine, esca complex, and environment: The disease triangle. Phytopathologia Mediterranea. 2019;**58**:17-37. DOI: 10.14601/Phytopathol_Mediterr-25086

[7] Kovács C, Balling P, Bihari Z, Nagy A, Sándor E. Incidence of grapevine trunk diseases is influenced by soil, topology and vineyard age, but not by *Diplodia seriata* infection rate in the Tokaj Wine Region Hungary. Phytoparasitica. 2017;**45**:21-32. DOI: 10.1007/s12600-017-0570-5

[8] Grigg D, Methven D, De Bei R, Rodríguez-López CM, Dry P, Collins C. Effect of vine age on vine performance of Shiraz in the Barossa valley, Australia: Effect of vine age on vine performance of Shiraz. Australian Journal of Grape and Wine Research. 2018;**24**:75-87. DOI: 10.1111/ajgw.12312

[9] Halleen F, Crous PW, Petrini O. Fungi associated with healthy grapevine cuttings in nurseries, with special reference to pathogens involved in the decline of young vines. Australasian Plant Pathology. 2003;**32**:47-52. DOI: 10.1071/AP02062

[10] Di Marco S, Osti F, Bossio D, Nocentini M, Cinelli T, Calzarano F, et al. Electrolyzed acid water: A clean technology active on fungal vascular pathogens in grapevine nurseries. Crop Protection. 2019;**119**:88-96. DOI: 10.1016/j.cropro.2019.01.005

[11] Mugnai L, Graniti A, Surico G. Esca (black measles) and brown wood-streaking: Two old and elusive diseases of grapevines. Plant Disease. 1999;**83**:404-418. DOI: 10.1094/PDIS.1999.83.5.404

[12] Graniti A, Surico G, Mugnai L. Esca of grapevine: A disease complex or a complex of diseases. Phytopathologia Mediterranea. 2000;**39**:16-20. DOI: 10.14601/Phytopathol_Mediterr-1539

[13] Surico G, Mugnai L, Marchi G. The esca disease complex. In: Ciancio A, Mukerji KG, editors. Integrated Management of Diseases Caused by Fungi, Phytoplasma and Bacteria. Dordrecht, Netherlands: Springer Science + Business Media B.V; 2008. p. 119

[14] Goufo P, Cortez I. A lipidomic analysis of leaves of Esca-affected grapevine suggests a role for galactolipids in the defense response and appearance of foliar symptoms. Biology. 2020;**9**(9):268. DOI: 10.3390/biology9090268

[15] van Niekerk JM, Bester W, Halleen F, Crous PW, Fourie PH. The distribution and symptomatology of grapevine trunk disease pathogens are influenced by climate. Phytopathologia Mediterranea. 2011;**50**:98-111. DOI: 10.14601/Phytopathol_Mediterr-1543

[16] Calzarano F, Osti F, Baránek M, Di Marco S. Rainfall and temperature influence expression of foliar symptoms of grapevine leaf stripe disease (esca complex) in vineyards. Phytopathologia Mediterranea. 2018;**57**:488-505. DOI: 10.14601/Phytopathol_Mediterr-23787

[17] Martín L, Fontaine F, Castaño FJ, Songy A, Roda R, Vallet J, et al. Specific profile of tempranillo grapevines related to Esca-leaf symptoms and climate conditions. Plant Physiology and Biochemistry. 2019;**135**:575-587. DOI: 10.1016/j.plaphy.2018.10.040

[18] Huber L, Gillespie TJ. Modeling leaf wetness in relation to plant disease epidemiology. Annual Review of Phytopathology. 1992;**30**:553-577. DOI: 10.1146/annurev.py.30.090192.003005

[19] McDowell N, Pockman WT, Allen CD, Breshears DD, Cobb N, Kolb T, et al. Mechanisms of plant survival and mortality during drought: Why do some plants survive while others succumb to drought. The New Phytologist. 2008;**178**:719-739. DOI: 10.1111/j.1469-8137.2008.02436.x

[20] Adams HD, Germino MJ, Breshears DD, Barron-Gafford GA, Guardiola-Claramonte M, Zou CB, et al. Nonstructural leaf carbohydrate dynamics of *Pinus edulis* during drought-induced tree mortality reveal role for carbon metabolism in mortality mechanism. The New Phytologist. 2013;**197**:1142-1151. DOI: 10.1111/nph.12102

[21] Fichot R, Laurans F, Monclus R, Moreau A, Pilate G, Brignolas F. Xylem anatomy correlates with gas exchange, water-use efficiency and growth performance under contrasting water regimes: Evidence from *Populus deltoides* x *Populus nigra* hybrids. Tree Physiology. 2009;**29**:1537-1549. DOI: 10.1093/treephys/tpp087

[22] Bauerle TL, Centinari M, Bauerle WL. Shifts in xylem vessel diameter and embolisms in grafted apple trees of differing rootstock growth potential in response to drought. Planta. 2011;**234**:1045-1054. DOI: 10.1007/s00425-011-1460-6

[23] Pouzoulet J, Pivovaroff AL, Santiago LS, Rolshausen PE. Can vessel dimension explain tolerance toward fungal vascular wilt diseases in woody plants. Lessons from Dutch elm disease and esca disease in grapevine. Frontiers in Plant Science. 2014;**5**:253. DOI: 10.3389/fpls.2014.00253

[24] Chiarappa L. Esca (black measles) of grapevine. An overview. Phytopathologia Mediterranean. 2000;**39**(1):11-15

[25] Calzarano F, Seghetti L, Del Carlo M, Cichelli A. Effect of Esca

on the quality of berries, musts and wines. Phytopathologia Mediterranea. 2004;**43**:125-135. DOI: 10.14601/Phytopathol_Mediterr-1729

[26] Mondello V, Songy A, Battiston E, Pinto C, Coppin C, Trotel-Aziz P, et al. Grapevine trunk diseases: A review of fifteen years of trials for their control with chemicals and biocontrol agents. Plant Disease. 2018;**102**:1189-1217. DOI: 10.1094/PDIS-08-17-1181-FE

[27] Lima MRM, Machado AF, Gubler WD. Metabolomic study of chardonnay grapevines double stressed with Esca-associated fungi and drought. Phytopathology. 2017;**107**(6):669-680. DOI: 10.1094/PHYTO-11-16-0410-R

[28] Crous PW, Grams W. «*Phaeomoniella chlamydospora*» Gen. et Comb. Nov., a causal organism of Petri grapevine decline and Esca. Phytopathologia Mediterranea. 2000;**39**:112-118. DOI: 10.14601/Phytopathol_Mediterr-1530

[29] Ciccarone C, Graniti A, Schiaffino A, Marras F. Molecular analysis of *Fomitiporia mediterranea* isolates from Esca-affected grapevines in southern Italy. Phytopathologia Mediterranea. 2004;**43**:268-272. DOI: 10.14601/Phytopathol_Mediterr-1748

[30] Fischer M, González-García V. An annotated checklist of European basidiomycetes related to white rot of grapevine (*Vitis vinifera*). Phytopathologia Mediterranea. 2015;**54**:281-298. DOI: 10.14601/Phytopathol_Mediterr-16293

[31] Fischer M. A new wood-decaying basidiomycete species associated with esca of grapevine: *Fomitiporia mediterranea* (Hymenochaetales). Mycological Progress. 2002;**1**:315-324. DOI: 10.1007/Fs11557-006-0029-4

[32] Fischer M, Kassemeyer HH. Fungi associated with esca disease of grapevine in Germany. Vitis. 2003;**42**:109-116. DOI: 10.5073/vitis.2003.42.109-116

[33] Fischer M. Biodiversity and geographic distribution of basidiomycetes causing Esca-associates white rot in grapevine: A worldwide perspective. Phytopathologia Mediterranea. 2006;**45**:30-42. DOI: 10.14601/Phytopathol_Mediterr-1846

[34] White CL, Halleen F, Mostert L. Symptoms and fungi associated with esca in South African vineyards. Phytopathologia Mediterranea. 2011;**50**:236-246. DOI: 10.14601/Phytopathol_Mediterr-8982

[35] Cloete M, Fischer M, Mostert L, Halleen F. Hymenochaetales associated with esca-related wood rots on grapevine with a special emphasis on the status of esca in South African vineyards. Phytopathologia Mediterranea. 2015;**54**:299-312. DOI: 10.14601/Phytopathol_Mediterr-16364

[36] Essakhi S, Mugnai L, Crous PW, Groenewald JZ, Surico G. Molecular and phenotypic characterisation of novel *Phaeoacremonium* species isolated from esca diseased grapevines. Persoonia. 2008;**21**:119-134. DOI: 10.3767/003158508X374385

[37] Spies C, Moyo P, Halleen F, Mostert L. *Phaeoacremonium* species diversity on woody hosts in the Western Cape Province of South Africa. Persoonia. 2018;**40**:26-62. DOI: 10.3767/persoonia.2018.40.02

[38] Borie B, Jacquiot L, Jamaux-Despréaux I, Larignon P, Péros JP. Genetic diversity in populations of the fungi *Phaeomoniella chlamydospora* and *Phaeoacremonium aleophilum*

on grapevine in France. Plant Pathology. 2002;**51**:85-96. DOI: 10.1046/j.0032-0862.2001.658.x

[39] Gramaje D, Mostert L, Groenewald JZ, Crous PW. *Phaeoacremonium*: From esca disease to phaeohyphomycosis. Fungal Biology. 2015;**119**:759-783. DOI: 10.1016/j.funbio.2015.06.004

[40] Fontaine F, Pinto C, Vallet J, Clément C, Gomes AC, Spagnolo A. The effects of grapevine trunk diseases (GTDs) on vine physiology. European Journal of Plant Pathology. 2016;**144**:707-721

[41] Hofstetter V, Buyck B, Croll D, Viret O, Couloux A, Gindro K. What if esca disease of grapevine were not a fungal disease. Fungal Diversity. 2012;**54**:51-67

[42] Guerin-Dubrana L, Fontaine F, Mugnai L. Grapevine trunk disease in European and Mediterranean vineyards: Occurrence, distribution and associated disease-affecting cultural factors. Phytopathologia Mediterranea. 2019;**58**:49-71

[43] Cortesi P, Fischer M, Milgroom MG. Identification and spread of *Fomitiporia punctata* associated with wood decay of grapevine showing symptoms of Esca. Phytopathology. 2000;**90**:967-972. DOI: 10.1094/PHYTO.2000.90.9.967

[44] Rego C, Oliveira H, Carvalho A, Phillips AJL. Involvement of *Phaeoacremonium spp*. and *Cylindrocarpon destructans* with grapevine decline in Portugal. Phytopathologia Mediterranea. 2000;**39**:76-79

[45] Armengol J, Vicent A, Torné L, Garcia-Figueres F, Garcia-Jimenez J. Fungi associated with esca and grapevine declines in Spain: A three-year survey. Phytopathologia Mediterranea. 2001;**2001**(40):325-329

[46] Redondo C, Tello ML, Avila A, Matea-Sagasta E. Spatial distribution of symptomatic grapevines with esca disease in the Madrid region (Spain). Phytopathologia Mediterranea. 2001;**40**:439-442

[47] Rumbos I, Rumbou A. Fungi associated with esca and young grapevine decline in Greece. Phytopathologia Mediterranea. 2001;**40**:330-335

[48] Rumbos I. Diseases and Pests of Grapevines. 6th ed. Athens, Greece: Stamoulis Publishing; 2016. p. 90

[49] Serra S, Mannoni MA, Ligios V. Studies on the susceptibility of pruning wounds to infection by fungi involved in grapevine wood diseases in Italy. Phytopathologia Mediterranea. 2008;**47**:234-246. DOI: 10.14601/Phytopathol_Mediterr-2727

[50] Moyo P, Allsopp E, Roets F, Mostert L, Halleen F. Arthropods vector grapevine trunk disease pathogens. Phytopathology. 2014;**104**:1063-1069. DOI: 10.1094/PHYTO-11-13-0303-R

[51] González-Domínguez E, Berlanas C, Gramaje D, Armengol J, Rossi V, Berbegal M. Temporal dispersal patterns of *Phaeomoniella chlamydospora*, causal agent of Petri disease and Esca, in Vineyards. Phytopathology. 2020;**110**(6):1216-1225. DOI: 10.1094/PHYTO-10-19-0400-R

[52] Halleen F, Baloyi MA, Bester MC, Mostert L. Aerial inoculum patterns of Petri disease pathogens in South African vineyards and rootstock mother blocks. Phytopathologia Mediterranea. 2020;**59**(3):515-536. DOI: 10.14601/Phyto-11370

[53] Larignon P, Dubos B. Preliminary studies on the biology of *Phaeoacremonium*. Phytopathologia Mediterranea. 2000;**39**:184-189. DOI: 10.14601/Phytopathol_Mediterr-1554

[54] Eskalen A, Gubler WD. Association of Spores of «*Phaeomoniella chlamydospora*», «*Phaeoacremonium inflatipes*», and «*Pm. aleophilum*» with grapevine cordons in California. Phytopathologia Mediterranea. 2001;**40**:429-432. DOI: 10.14601/Phytopathol_Mediterr-1613

[55] Edwards J, Laukart N, Pascoe IG. In situ sporulation of *Phaeomoniella chlamydospora* in the vineyard. Phytopathologia Mediterranea. 2001;**40**:61-66

[56] Rooney-Latham S, Eskalen A, Gubler WD. Occurrence of *Togninia minima* perithecia in Esca-affected vineyards in California. Plant Disease. 2005;**89**:867-871. DOI: 10.1094/PD-89-0867

[57] Whiteman SA, Stewart A, Ridgway HJ, Jaspers MV. Infection of rootstock mother-vines by *Phaeomoniella chlamydospora* results in infected young grapevines. Australasian Plant Pathology. 2007;**36**:198-203. DOI: 10.1071/AP07012

[58] Andreini L, Caruso G, Bertolla C, Scalabrelli G, Viti R, Gucci R. Gas exchange, stem water potential and xylem flux on some grapevine cultivars affected by Esca disease. South African Journal of Enology and Viticulture. 2009;**30**:142-147. DOI: 10.21548/30-2-1434

[59] Markakis E, Koubouris G, Sergentani C, Ligoxigakis E. Evaluation of Greek grapevine cultivars for resistance to *Phaeomoniella chlamydospora*. European Journal of Plant Pathology. 2017;**149**:277-283. DOI: 10.1007/s10658-017-1186-9

[60] Martínez-Diz M, Díaz-Losada E, Barajas E, Ruano-Rosa D, Andrés-Sodupe M, Gramaje D. Screening of Spanish Vitis vinifera germplasm for resistance to *Phaeomoniella chlamydospora*. Scientia Horticulturae. 2019;**246**:104-109. DOI: 10.1016/j.scienta.2018.10.049

[61] Marchi G. Susceptibility to Esca of various grapevine (*Vitis vinifera*) cultivars grafted on different rootstock in a vineyard in the Province of Siena (Italy). Phytopathologia Mediterranea. 2001;**40**:27-36. DOI: 10.14601/Phytopathol_Mediterr-1589

[62] Sparapano L, Bruno G, Ciccarone C, Graniti A. Infection of grapevines by some fungi associated with Esca. II. Interaction among «*Phaeoacremonium chlamydosporum*, *P. aleophilum*» and «*Fomitiporia punctata*». Phytopathologia Mediterranea. 2000;**39**:53-58. DOI: 10.14601/Phytopathol_Mediterr-1543

[63] Feliciano A, Eskalen A, Gubler WD. Differential susceptibility of three grapevine cultivars to *Phaeoacremonium aleophilum* and *Phaeomoniella chlamydospora* in California. Phytopathologia Mediterranea. 2004;**43**:66-69. DOI: 10.14601/Phytopathol_Mediterr-1727

[64] Larignon P, Fontaine F, Farine S, Clément C. Esca et black dead arm: Deux acteurs majeurs des maladies du bois chez la vigne. Comptes Rendus Biologies. 2009;**332**:765-783

[65] Lorrain B, Ky I, Pasquier G, Jourdes M, Dubrana LG, Gény L, et al. Effect of Esca disease on the phenolic and sensory attributes of Cabernet Sauvignon grapes, musts and wines.

Australian Journal of Grape and Wine Research. 2012;**18**:64-72. DOI: 10.1111/j.1755-0238.2011.00172.x

[66] Zabalgogeazcoa I. Fungal endophytes and their interaction with plant pathogens: A review. Spanish Journal of Agricultural Research. 2008;**6**:138-146. DOI: 10.5424/sjar/200806S1-3822006

[67] Amponsah NT, Jones EE, Ridgway HJ, Jaspers MV. Identification, potential inoculum sources and pathogenicity of botryosphaeriaceous species associated with grapevine dieback disease in New Zealand. European Journal of Plant Pathology. 2011;**131**:467-482. DOI: 10.1007/s10658-011-9823-1

[68] Verhoeff K. Latent Infections by Fungi. Annual Review of Phytopathology. 1974;**12**:99-110. DOI: 10.1146/annurev.py.12.090174.000531

[69] Petit AN, Vaillant N, Boulay M, Clément C, Fontaine F. Alteration of photosynthesis in grapevines affected by Esca. Phytopathology. 2006;**96**:1060-1066. DOI: 10.1094/PHYTO-96-1060

[70] Calzarano F, Di Marco S, Cesari A. Benefit of fungicide treatment after trunk renewal of vines with different types of esca necrosis. Phytopathologia Mediterranea. 2004;**43**:116-124

[71] Magnin-Robert M, Letousey P, Spagnolo A, Rabenoelina F, Jacquens L, Mercier L, et al. Leaf stripe form of esca induces alteration of photosynthesis and defence reactions in presymptomatic leaves. Functional Plant Biology. 2011;**38**(11):856-866. DOI: 10.1071/FP11083

[72] Letousey P, Baillieul F, Perrot G, Rabenoelina F, Boulay M, Vaillant-Gaveau N, et al. Early events prior to visual symptoms in the apoplectic form of grapevine esca disease. Phytopathology. 2010;**100**:424-431. DOI: 10.1094/PHYTO-100-5-0424

[73] Eskalen A, Feliciano AJ, Gubler WD. Susceptibility of grapevine pruning wounds and symptom development in response to infection by *Phaeoacremonium aleophilum* and *Phaeomoniella chlamydospora*. Plant Disease. 2007;**91**:1100-1104. DOI: 10.1094/PDIS-91-9-1100

[74] Fischer M, Kassemeyer HH. Water regime and its possible impact on expression of Esca symptoms in *Vitis vinifera*: Growth characters and symptoms in the greenhouse after artificial infection with *Phaeomoniella chlamydospora*. Vitis. 2012;**51**:129-135. DOI: 10.5073/vitis.2012.51.129-135

[75] Lovisolo C, Schubert A. Effects of water stress on vessel size and xylem hydraulic conductivity in *Vitis vinifera* L. Journal of Experimental Botany. 1998;**49**:693-700. DOI: 10.1093/jxb/49.321.693

[76] Pasquier G, Lapaillerie D, Vilain S, Dupuy JW, Lomenech AM, Claverol S, et al. Impact of foliar symptoms of "Esca proper" on proteins related to defence and oxidative stress of grape skins during ripening. Proteomics. 2013;**13**:108-118. DOI: 10.1002/pmic.201200194

[77] Sosnowski M, Luque J, Loschiavo A, Martos S, Garcia-Figueres F, Wicks T, et al. Studies on the effect of water and temperature stress on grapevines inoculated with *Eutypa lata*. Phytopathologia Mediterranea. 2011;**50**:127-138. DOI: 10.14601/Phytopathol_Mediterr-8959

[78] Retief E, McLeod A, Fourie PH. Potential inoculum sources of *Phaeomoniella chlamydospora* in

South African grapevine nurseries. European Journal of Plant Pathology. 2006;**115**:331-339

[79] Whiteman SA, Jaspers MV, Stewart A, Ridgway HJ. Identification of potential sources of *Phaeomoniella chlamydospora* in the grapevine propagation process. In: 8th International Congress of Plant Pathology 2-7 February 2003; Christchurch, New Zealand. 2003. p. 86

[80] Edwards J, Pascoe IG, Salib S, Laukart N. Hot treatment of grapevine cuttings reduces incidence of *Phaeomoniella chlamydospora* in young vines. Phytopathologia Mediterranea. 2004;**43**:158-159

[81] Fourie PH, Halleen F. Investigation on the occurrence of *Phaeomoniella chlamydospora* in canes of rootstock mother vines. Australasian Plant Pathology. 2002;**31**:425-427. DOI: 10.1071/AP02049

[82] Aroca A, García-Figueres F, Bracamonte L, Luque J, Raposo R. A survey of trunk disease pathogens within rootstocks of grapevines in Spain. European Journal of Plant Pathology. 2006;**115**:195-202

[83] Giménez-Jaime A, Aroca A, Raposo R, García-Jiménez J, Armengol J. Occurrence of fungal pathogens associated with grapevine nurseries and the decline of young vines in Spain. Journal of Phytopathology. 2006;**154**:598-602

[84] Whiteman S, Jaspers M, Stewart A, Ridgway H. Soil as a source of *Phaeomoniella chlamydospora* (causal agent of Petri disease). In: Abstracts of the 4th International Workshop on Grapevine Trunk Diseases; 20-21 January 2005. Stellenbosch, South Africa: University of Stellenbosch; 2005. p. 78

[85] Retief E, Damm U, vanNiekerk JM, McLeod A, Fourie PH. A protocol for molecular detection of *Phaeomoniella chlamydospora* in grapevine wood. South African Journal of Science. 2005;**101**:139-142

[86] Damm U, Fourie PH. Development of a cost-effective protocol for molecular detection of fungal pathogens in soil. South African Journal of Science. 2005;**101**:135-139

[87] Soll DR. The ins and outs of DNA fingerprinting the infectious fungi. Clinical Microbiology Reviews. 2000;**13**:332-370

[88] Pilotti M, Gervasi F, Brunetti A. Molecular identification of *Fomitiporia mediterranea* and *Eutypa lata/Libertella blepharis* in *Platanus × acerifolia*. Journal of Phytopathology. 2005;**153**:193-202

[89] Pollastro S, Dongiovanni C, Abbatecola A, Faretra F. Observations on the fungi associated with esca and on spatial distribution of esca-symptomatic plants in Apulian (Italy) vineyards. Phytopathologia Mediterranea. 2000;**39**:206-210

[90] Jamaux-Despréaux I, Péros JP. Genetic structure in populations of the fungus *Fomitiporia punctata* associated with the esca syndrome in grapevine. Vitis. 2003;**42**:43-51

[91] Lupo S, Bettucci L, Pérez A, Martínez S, Césari C, Escoriaza G, et al. Characterization and identification of the basidiomycetous fungus associated with 'hoja de malvón' grapevine disease in Argentina. Phytopathologia Mediterranea. 2006;**45**:110-116

[92] Pottinger B, Stewart A, Carpenter M, Ridgway HJ. Low genetic variation detected in New Zealand populations of *Phaeomoniella chlamydospora*.

Phytopathologia Mediterranea. 2002;**41**:199-211

[93] Blanco-Ulate B, Rolshausen PE, Cantu D. Draft genome sequence of the grapevine dieback fungus *Eutypa lata* UCR-EL1. Genome Announcements. 2013;**1**:e00228-e00213

[94] Antonielli V, Compant S, Strauss J, Sessitsch A, Berger H. Draft genome sequence of *Phaeomoniella chlamydospora* strain RR-HG1, a grapevine trunk disease (Esca)-related member of the ascomycota. Genome Announcements. 2014;**2**:e00098-e00014

[95] Bendel N, Backhaus A, Kicherer A, Köckerling J, Maixner M, Jarausch B, et al. Detection of two different grapevine yellows in *vitis vinifera* using hyperspectral imaging. Remote Sensing. 2020;**12**:4151. DOI: 10.3390/rs12244151

[96] Humplík JF, Lazár D, Husičková A, Spíchal L. Automated phenotyping of plant shoots using imaging methods for analysis of plant stress responses—A review. Plant Methods. 2015;**11**:29. DOI: 10.1186/s13007-015-0072-8

[97] Rançon F, Bombrun L, Keresztes B, Germain C. Comparison of SIFT encoded and deep learning features for the classification and detection of esca disease in Bordeaux vineyards. Remote Sensing. 2018;**11**(1):1-26

[98] Waite H, May P. The effects of hot water treatment, hydration and order of nursery operations on cuttings of Vitis vinifera cultivars. Phytopathologia Mediterranea. 2005;**44**:144-152

[99] Gramaje D, Aroca A, Raposo R, García-Jiménez J, Armengol J. Evaluation of fungicides to control Petri disease pathogens in the grapevine propagation process. Crop Protection. 2009;**28**:1091-1097

[100] Fourie PH, Halleen F. Proactive control of Petri disease of grapevine through treatment of propagation material. Plant Disease. 2004;**88**:1241-1245

[101] Habib W, Pichierri A, Masiello N, Pollastro S, Faretra F. Application of hot water treatment to control *Phaeomoniella chlamydospora* in grapevine plant propagation materials. Phytopathologia Mediterranea. 2009;**48**:186

[102] Caudwell A, Larrue J, Boudon-Padieu E, Mclean GD. *Flavescence dorée* elimination from dormant wood of grapevines by hot-water treatment. Australian Journal of Grape and Wine Research. 1997;**3**:21-25

[103] Laukart N, Edwards J, Pascoe IG, Nguyen NK. Curative treatments trialled on young grapevines infected with *Phaeomoniella chlamydospora*. Phytopathologia Mediterranea. 2001;**40**:459-463

[104] Waite H, Morton L. Hot water treatment, trunk diseases and other critical factors in the production of high-quality grapevine planting material. Phytopathologia Mediterranea. 2007;**46**:5-17

[105] Crocker J, Waite H, Wright P, Fletcher G. Source area management: Avoiding cutting dehydration and good nursery management may be the keys to successful hot water treatment. The Australian and New Zealand Grapegrower and Winemaker. 2002;**461**:33-37

[106] Halleen F, Fourie PH. An integrated strategy for the proactive management of grapevine trunk disease pathogen infections in grapevine nurseries. South African Journal of Enology and Viticulture. 2016;**37**:104-114

[107] Halleen F, Mostert L, Crous PW. Pathogenicity testing of lesser-known

vascular fungi of grapevines. Australasian Plant Pathology. 2007;**36**:277-285

[108] Di Marco S, Metruccio EG, Moretti S, Nocentini M, Carella G, Pacetti A, et al. Activity of *trichoderma asperellum* strain ICC 012 and *trichoderma gamsii* strain ICC 080 toward diseases of Esca complex and associated pathogens. Frontiers in Microbiology. 2022;**12**:813410. DOI: 10.3389/fmicb.2021.813410

[109] Pertot I, Prodorutti D, Colombini A, Pasini L. *Trichoderma atroviride* SC1 prevents *Phaeomoniella chlamydospora* and *Phaeoacremonium aleophilum* infection of grapevine plants during the grafting process in nurseries. BioControl. 2016;**61**:257-267

[110] Elena G, Luque J. Seasonal susceptibility of grapevine pruning wounds and cane colonization in Catalonia, Spain following artificial infection with *Diplodia seriata* and *Phaeomoniella chlamydospora*. Plant Disease. 2016;**100**:1651-1659

[111] Mugnai L. Editor's note and dedication. Phytopathologia Mediterranea. 2011;**50**(4):3-4

[112] Mostert L, Halleen F, Fourie PH, Crous PW. A review of *Phaeoacremonium* species involved in Petri disease and esca of grapevines. Phytopathologia Mediterranea. 2006;**45**:12-29

[113] Úrbez-Torres JR, Bruez E, Hurtado J, Gubler WD. Effect of temperature on conidial germination of Botryosphaeriaceae species infecting grapevines. Plant Disease. 2010;**94**:1476-1484

[114] van Niekerk JM, Calitz FJ, Halleen F, Fourie PH. Temporal spore dispersal patterns of grapevine trunk pathogens in South Africa. European Journal of Plant Pathology. 2010;**127**:375-390

[115] Munkvold GP, Marois JJ. Factors associated with variation in susceptibility of grapevine pruning wounds to infection by *Eutypa lata*. Phytopathology. 1995;**85**:249-256

[116] Weber EA, Trouillas F, Gubler W. Double pruning of grapevines: A cultural practice to reduce infections by *Eutypa lata*. American Journal of Enology and Viticulture. 2007;**58**:61-66

[117] Rolshausen PE, Úrbez-Torres JR, Rooney-Latham S, Eskalen A, Smith RJ, Gubler WD. Evaluation of pruning wound susceptibility and protection against fungi associated with grapevine trunk diseases. American Journal of Enology and Viticulture. 2010;**61**:113-119

[118] Herche R. Control strategies for trunk diseases of grapevine (*Vitis vinifera* L.) [MSc thesis]. Davis, CA, USA: University of California Davis; 2009

[119] Aloi C, Bigot G, Bortolotti P, Cotromino M, Di Marco S, Faccini F. Six years of trials on the activity of remedier® against Esca disease complex in young and mature Italian vineyards. Phytopathologia Mediterranea. 2017;**56**:566

[120] Pacetti A, Moretti S, Pinto C, Farine S, Bertsch C, Mugnai L. Trunk surgery as a tool to reduce foliar symptoms in diseases of the esca complex and its influence on vine wood microbiota. Journal of Fungi. 2021;**7**:521

[121] Thibault M. Les maladies du bois de la vigne-Le Curetage. In: Proceedings of the Service Interprofessionnel de Conseil Agronomique, de Vinification et d'Analyses du Centre, 17-18 November 2015. Colmar, France: Huniversité de Haute-Alsace; 2015. pp. 55-57

[122] Smart R. Trunk diseases: Timely trunk renewal to overcome trunk disease. Wine and Viticulture Journal. 2015;**30**(5):44

[123] Mondello V, Giambra S, Conigliaro G, Francesca N, Burruano S. Fungal pathogens associated with grapevine trunk diseases in young vineyards in Sicily. Phytopathologia Mediterranea. 2020;**59**:453-463

[124] Darrieutort G, Lecomte P. Evaluation of a trunk injection technique to control grapevine wood diseases. Phytopathologia Mediterranea. 2007;**46**:50-57

[125] Del Frari G, Gobbi A, Aggerbeck MR, Oliveira H, Hansen LH, Ferreira RB. Characterization of the wood mycobiome of *Vitis vinifera* in a vineyard affected by esca. Spatial distribution of fungal communities and their putative relation with leaf symptoms. Frontiers in Plant Science. 2019;**10**:910

[126] Machado TO, Beckers SJ, Fischer J, Müller B, Sayer C, de Araújo PHH. Bio-based lignin nanocarriers loaded with fungicides as a versatile platform for drug delivery in plants. Biomacromolecules. 2020;**21**:2755-2763

[127] Jourdes M. Impact de l'esca sur la qualité du raisin et du vin. VITI, de la vigne au vin [Internet]. 2016. Available from: https://www.mon-viti.com/node/21452/activer

[128] Calzarano F, Cichelli A, Odoardi M. Preliminary evaluation of variations in composition induced by esca on cv. Trebbiano d'Abruzzo grapes and wines. Phytopathologia Mediterranea. 2001;**40**:443-448

Chapter 4

Enzymes and Biochemical Catalysis in Enology: Classification, Properties, and Use in Wine Production

Elena Cristina Scutarașu, Camelia Elena Luchian, Lucia Cintia Colibaba and Valeriu Cotea

Abstract

The quality of wine, its structure, and its chemical composition are dependent on the grapes' characteristics as raw material, alcoholic fermentation particularities, and the applied oenological practices. Awareness of the significant role that enzymes play in winemaking contributes to the development of different new strategies for optimizing the production process. Numerous studies confirmed the positive impact of using enzymes in food and beverage industries, in improving the quality of final products, and optimization of applied production technologies. This chapter aims to present the link between biochemical processes that involve enzymes and the quality of wine as a final food product.

Keywords: wine, biochemical catalysts, enzymes activity, food processing, optimization processes

1. Introduction

Enzymes are functional units of cellular metabolism that catalyze biological reactions. In other words, they lend protein compounds a catalytic role in accelerating the transformation of chemicals into living organisms without being expended during the reaction. The converted substance is the substrate of action, while the compound resulting from the enzymatic activity is called the reaction product [1–4]. Enzymes can speed up the reaction in cells up to 10^{16} times. As such, the presence of a relatively small amount of enzymes can catalyze the bioconversion of a large amount of substrate [3]. Even though enzymes are formed inside living cells, they can have *in vitro* activity (e.g., various enzymes in blood plasma), and they are also present in industrial processes. Enzymatic processes have been known since antiquity, with enzymes being initially used under the name of ferments in correlation with their role in the fermentation of sugars and subsequent transformation into alcohol. The earliest reference to the commercial use of enzymes is found in a description of wine in Hammurabi's Code (ancient Babylon, about 2100 BC). Ancient texts of the early civilizations of Rome, Greece, Egypt, and China also contain a number of references to the technological process of vinegar, which is based on the enzymatic conversion of alcohol to acetic acid. Today, these compounds continue to play a key role in many food and beverage manufacturing processes, as well as in non-food products (e.g., laundry detergents that dissolve stains using proteolytic enzymes). The analysis and action of enzymes have caught the attention of scientific researchers not only as a focus of scholarly interest, but also because of their many practical needs for society [4]. Much of the research in biochemistry is devoted to analyzing the activity of enzymes. The first theory of chemical catalysis put forth by Berzelius, referred to the hydrolysis of starch, a reaction catalyzed rather by diastase (amylase) than by mineral acids. Thus, the presence of enzymes as biological catalysts specific to living organisms can explain many biological processes, such as fermentation or digestion. In a follow-up to Réanmur's studies, Spallanzani demonstrated the role of the enzymes found in gastric juice in the process of digestion. In 1836, Schwann coined the name of the gastric juice enzyme known as pepsin, while the name trypsin, an enzyme present in gastric juice, was coined by Kühne. In 1897, Eduard Buchner extracted from yeast cells the enzymes involved in the catalysis of alcoholic fermentation, which function independently of cellular structure [5]. In 1870, the Danish chemist Hansen managed to extract renin from the stomach of calves, which significantly improved both the quality and the quantity of cheese production. In 1921, Fleming discovered lysozyme, a component of tears, saliva, leukocytes, skin, nails, and human milk, which is widely spread in both animals and plants. He published the first articles on the subject between 1922 and 1927. In 1926, James Sumner managed to isolate urease from the jack bean (*Canavalia ensiformis*), the first pure crystalline enzyme. His observations were of particular importance for the development of enzymology, confirming the protein structure of enzymes. The name enzyme was given by Kühne (1867), while Stern was the first to observe the first enzyme-substrate complex in 1935 [5, 6]. Steady advances over time had a major impact on the enzyme industry, such as the production and marketing of glucoamylase, which catalyzes the production of glucose from starch with superior efficiency compared to the chemical procedure of acid hydrolysis. Consequently, the launch of enzyme-based detergents was made possible [3].

As to the beverage industry, enzymes were first used in the 1930s to make wine and fruit juices, with Boidin and Effront discovering bacterial amylase. The fruit drinks and juices industry began using pectinase in the late 1940s to improve clarification and filtration. These types of enzymatic preparations also began to be tested in the oenological sector. After 1974, they were officially authorized in the oenological industry by the Ciba-Geigy Company, with Ultrazym 100 as the first enzymatic preparation proposed. It was only in the 1980s that β-glucanases were authorized, which helped to solve the problems of clarifying and filtering wines obtained from botrytized grapes. In the late 1980s, β-glycosidase-enriched pectolytic enzymatic preparations began to appear on the market following a close collaboration between French (Montpellier) and Australian (Australian Wine Research Institute) researchers, and Gist Brocades [6]. Enzymes offer flexible, high-performance solutions that ensure high-quality products boasting a higher nutritional intake, cost-effectiveness, and guaranteed consumer satisfaction. Food enzymes are cost-effective and provide reliable food security, which explains why they are increasingly in demand in the food industry. Understanding the crucial role that these oenological products have in winemaking contributes to the development of optimizing strategies in improving the structure, chemical composition of the final product, and implicitly their sensory profiles [1, 3].

Grape pomace, the main by-product of the wine industry, has been shown to be an important source of nutrients. Botella *et al.* [7] studied the production of hydrolytic enzymes from grape remnants (cellulases, xylanases, and pectinases) under the influence of *Aspergillus awamori*. The volume of industrial use of enzymes has gone up recently due to their many advantages. Being of natural origin, enzymes have no toxicity and show a negligible impact on the environment. These catalysts present the specificity of action insofar as they are selective with regard to both the substrate they act upon and the catalyzed reaction. Enzymes act effectively in moderate temperature conditions, show quick action at relatively low concentrations and the reaction rate can be easily controlled by adjusting temperature, pH, and quantity. Moreover, enzymes can be easily inactivated once the reactions have produced the desired result. These preparations are considered technological aids and are not found in the final product [2].

2. Name and classification of enzymes

To date, more than 6000 different types of enzymes are known. The classification and naming of enzymes are generally based on the type and mechanism of the reaction they catalyze. The classification of enzymes is based on the principles established by the Enzyme Commission of the International Union of Biochemistry. The following criteria are laid out:

1. Enzymes and the reactions they catalyze are divided into six different classes, each in turn divided into subclasses.

2. The name of the enzyme provides information on the name of the substrate and the type of catalyzed reaction, followed by the suffix -ase, except for proteolytic enzymes, where the suffix is -in (trypsin). For example, protein hydrolysis is catalyzed by proteases.

3. For a correct and positive identification, each enzyme is assigned a 4-digit code number, as follows: the first digit refers to the reaction class it belongs to; the second and third digits indicate the subclass and sub-subclass; the fourth is the serial number of the enzyme in the subclass [2, 4, 6, 8].

Enzymes are classified as follows:

Oxidoreductases (EC 1). Enzymes that catalyze redox reactions belong to this class, with the oxidized substrate as hydrogen-donor. Put differently, oxidoreductases catalyze the transfer of hydrogen, oxygen, or electron atoms from one substrate to another [4, 6, 9]. The systematic name is based on the donor-acceptor oxidoreductase: dehydrogenase or reductase. The recommended name is made up of the donor's name and the endings: dehydrogenase, oxidase, reductase, oxygenase, and peroxidase. The name oxidase is used only in cases where O_2 is the acceptor, while that of oxygenase is used when part of the O_2 molecule is framed in the corresponding substrate. Oxidoreductases make up about 25% of all enzymes [10].

Transferases (EC 2) are enzymes that transfer a group, for example, a methyl group or a hydroxyl glycosidic group, from one compound (generally considered a donor) to another (acceptor). The systematic names are made up according to the donor of the scheme: donor-acceptor or group transferase. Recommended common names are *group transferase donor* or *group transferase acceptor*, but the name *group kinase acceptor* is also accepted for some phosphotransferases (e.g., hexokinase and glucokinase). This class makes up about 30% of all known enzymes. Of these, of enological interest are reactions involving the transfer of phosphoric acid residue (H_3PO_4), carbonate ion ($-(CO_3)^{-2}$), carbon dioxide (CO_2), water (H_2O), ammonia (NH_3), and amino groups ($-NH_2$) [9, 10].

Hydrolases (EC 3) amount to 24% of all known enzymes to date and catalyze hydrolytic reactions. These are group transfer reactions where the acceptor is always water, the systematic name is the hydrolase substrate (e.g., peptidyl-peptide hydrolase), and the common name *-ase* substrate (e.g., methylesterase and *o*-glycosidase). Of interest in the food industry are α-amylases (EC. 3.2.1.1), β-amylases (EC. 3.2.1.2), lactase (EC. 3.2.1.23), lipase (EC. 3.1.1.3), proteases - amino peptidases (EC. 3.4.11), trypsin (EC.3.4.21.4), subtilisin (EC. 3.4.21.62), papain (EC. 3.4.22.2), ficin (EC. 3.4.22.3), pepsin (EC. 3.4. 23.1), and chymosin (EC. 3.4.23.4). The hydrolases commonly found in must and wine are of fungal origin and come either from the plant's microflora or from an external source following administration of treatments with enzymatic preparations [4, 9, 10].

Lyases (EC 4) represent 13% of the overall number of enzymes known. They catalyze the addition or removal of non-hydrolytic groups from the structure of the substrate, with the formation of double bonds or acyclic structures of the type C–C, C–O, and C–N. The systematic name consists of substrate group—lyases. The (common) historical names are created according to the group removed, namely: decarboxylase, when carbon dioxide is removed; dehydratase, when water is removed; aldolase, when the deleted group is of the aldehyde type. Out of the total lyases carbon-carbon lyases, carbon-nitrogen lyases, and carbon-oxygenates are most significant for the wine industry [4, 9, 10].

Isomerases (EC 5) catalyze isomerization reactions, as a result of which a molecule is converted from one isomer to another. They amount to 3% of all known enzymes. In general, the systematic name corresponds to the traditional one and is formed by: *substrate isomerization type class name*. Depending on the type of isomerism, the enzymes in this class can be divided into subclasses, namely: isomerases, racemases, *cis*- and *trans*-isomerases, epimerases, mutases, tautomerases, or cycloisomerases. Of these, the following are important in the enology field: lactate racemase, glucose isomerase, glucose phosphate isomerase, carotenoid isomerase, and triosephosphate isomerase [10].

Ligases (EC 6) are enzymes that catalyze the binding of two molecules coupled with the hydrolysis of a pyrophosphate bond to ATP or a similar triphosphate. Enzymes in this class are involved in condensation reactions. The systematic name is A: B ligase form (ADP-forming), where A and B are the two substrates, followed by the class name and the product resulting from hydrolysis. Ligases represent 5% of all known enzymes. Also called synthetases, they have the property of catalyzing bimolecular reactions to form new carbon-carbon or carbon-heteroatom bonds. Of these, those of interest in the enology industry are asparagine synthetase, acetyl-coA synthetase, succinyl-coA synthetase, glutamine synthetase, glutathione synthetase, and pyruvate carboxylase [6, 9].

3. Enzyme structure and reaction mechanism

Enzymes are macromolecular organic substances of protein origin that are spherical in shape and have primary, secondary, tertiary, and quaternary-type structures. They can be classified into holoproteins and heteroproteins. The catalytic properties of enzymes are generated by the spatial structural configuration of the molecule that participates in the development of enzymatic activity and by the existence of catalytic sites in their molecule to which the substrate, activator, or inhibitor, will bind, as appropriate [2]. There are also exceptions, such as the ribonuclease P, defined as a protein-ribonucleic acid complex whose enzymatic activity is not due to the protein, but to the ribonucleic acid [11]. In terms of chemical structure, enzymes are simple proteins (mono-components), such as trypsin, pepsin, lipase, sucrose, amylase, or conjugates (bicomponents), that is, apoenzymes and coenzymes. Some coenzymes are derivatives of vitamins (NAD, TDP, coenzyme A, etc.) or metal ions (Zn, Mn, Ca, etc.). Apoenzymes and coenzymes show a synergistic action by only acting in tandem. The apoenzyme is the protein macromolecule responsible for the specificity of the reaction, while the coenzyme triggers the reaction, both forming a fermentative complex [10]. The composition of the apoenzyme includes the catalytic site (a distinct area framed by a group of amino acids, which is set apart from the rest of the amino acids by their function, in which the specific reaction substrates bind) and the allosteric site (a distinct area for the activator or inhibitor to bind). Enzymes with both catalytic and regulatory functions are called allosteric enzymes [6]. The mechanism of enzymatic action has been explained by many researchers. Thus, Ogston [12] reported that at a wavelength of 280 nm, three points of interaction between enzyme and substrate were identified, which explains the phenomenon of stereospecificity of enzymes. These interactions have either a binding or a catalytic function. Binding sites (active sites) connect to specific groups in the substrate to ensure a stable orientation of the enzyme and substrate molecules with the reaction group in the vicinity of the catalytic sites. The three-point interaction theory cannot explain thoroughly the action and specificity of the enzyme, and there are other hypotheses in this regard. The action of enzymes is considered to occur in two stages, as shown in **Figure 1**: the active site of the enzyme initially combines with the substrate to form an enzyme-substrate complex (ES); the latter then decomposes to form the products (P) and the free enzyme (E), which can react yet again [9].

For the reaction to take place, the reacting molecules (substrate) require a certain amount of energy (activation) to traverse the transition state of the reaction and turn into reaction products. In 1888, the catalytic action of enzymes was explained by the Swedish chemist Svante Arrhenius, who proposed that the substrate and enzyme are combined to form an intermediate compound known as the enzyme-substrate complex

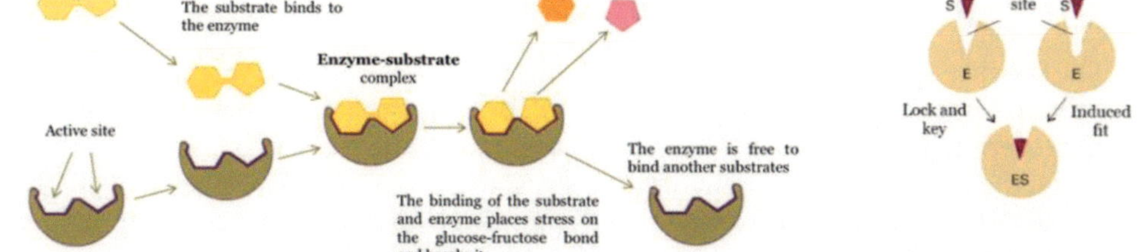

Figure 1.
Schematic representation of the enzymes' action (left) and illustration of the lock-and-key and induced fit model (right) [13].

(ES). This complex decomposes into a reaction product (P) and an active enzyme (E). The total enzyme-catalyzed reaction can be represented as: S + E → ES → P + E.

In general, enzyme-catalyzed reactions cover the following stages:

a. The substrate molecule comes into contact with the active site of the enzyme through non-covalent bonds. The active site is the area of the enzyme that combines with the substrate.

b. The substrate and the enzyme form an enzyme-substrate complex.

c. The substrate molecule is transformed into a reaction product either by rearranging the atoms or by decomposing the substrate or combining it with another molecule.

d. Dissolution of the ES complex leads to the formation of the reaction product, which is released by the active site of the enzyme.

e. The nature of the enzyme is unchanged and can catalyze a new reaction [6].

The mechanisms of enzymatic action are commonly explained by two proposed models:

1. **The lock-and-key hypothesis.** In 1894, Fischer put forth his theory by suggesting that both substrate and enzyme have specific geometric shapes that match. The hypothesis specifies that the active site of an enzyme has a unique configuration that is complementary to the substrate structure (key), and therefore allows the two molecules to match [6]. According to this model, the structures show rigidity by remaining fixed throughout the binding process [2, 9].

2. **The induced fit hypothesis.** In 1958, Koshland proposed some changes to the lock-and-key hypothesis detailed above by positing that the essential functional groups on the active site of the free enzyme are not in their optimal positions for catalysis. Because enzymes are so flexible, when the substrate molecule binds to them, the active site of the enzymes takes on a favorable geometric shape to reach the transition state. As per Koshland's suggestion, the substrate induces a configuration change in the enzymes that aligns the amino acid residues or other groups so as to bind and catalyze the substrate [2, 6].

4. Enzyme specificity

The specificity of the enzymes can be exhibited either on the substrate or on the reaction. In other words, an enzyme has an affinity for the substrate it acts on and for the reaction it catalyzes [11]. Enzymes have varying degrees of specificity. For example, the enzyme alcohol dehydrogenase catalyzes the dehydrogenation of high-efficiency ethanol and low-efficiency methanol. Such an enzyme is seen as specific to a compound, and not to a class of substances. Moreover, with regard to the reaction specifics, an enzyme can only catalyze a transformation of the substrate. For example, L-amino acid oxidase catalyzes the oxidation of L-amino acids to produce the corresponding keto-acids, ammonia, and hydrogen peroxide. However, the racemization of

L-amino acids into D-amino acids is catalyzed by an enzyme other than L-amino acid oxidase, that is, amino acid racemase [11]. To act as a catalyst, most enzymes need a molecule known as a cofactor, which is a non-protein chemical compound bound to an inactive protein part of the enzyme (apoenzyme) to increase its biological activity [4, 6]. The active complex of the apoenzyme (protein part) together with the cofactor (coenzyme/prosthetic group) constitutes the holoenzyme. Two categories of cofactors are known, namely: coenzymes and prosthetic groups. The cofactor may be a metal ion and/or an organic molecule. As a specific type of cofactor, coenzymes are organic molecules that bind to enzymes to ensure their functioning. Many coenzymes are derived from vitamins. Prosthetic groups are also cofactors that often bind closely to proteins or enzymes through a covalent bond [6]. Enzymes have the ability to catalyze biochemical reactions in cells, something specific to catabolic and anabolic processes. For example, the enzymes involved in photosynthesis are located in the chloroplasts, those of the glycolytic cycle are found in the cytoplasm, the enzymes of the Krebs cycle are present in the mitochondria, etc. Moreover, the intensity of the physiological process depends on the activity of the involved enzymes. For example, phosphofructokinase is involved in the phosphorylation reaction of fructose in plants, causing biodegradation of hexoses in the glycolytic cycle; phenylalanine ammonia lyase plays a role in the biosynthesis of phenols, anthocyanins, lignins; chlorophylls are responsible for catalyzing the chlorophyll decomposition reaction; polyphenol oxidases are involved in the catalysis of oxidation reactions of polyphenols, etc. [11].

5. Enzyme solubility

Enzymes are globular proteins soluble in aqueous solvents or dilute saline solutions. Their solubility increases through weak ionic interactions, such as hydrogen bonds. Some of the factors that influence or interfere with this process and have an effect on the solubility of enzymes are salt concentration, pH, temperature, and solvent structure. Solubility can be increased by adding neutral salt in low concentrations. When using salts with a higher solubility, such as ammonium sulfate, some proteins will precipitate only in certain concentrations. Most proteins will precipitate at more than 80% $(NH_4)_2SO_4$. Cations such as Zn^{2+} and Pb^{2+} decrease the solubility of enzymes to form insoluble complexes with the enzymatic protein. Proteins are also precipitated by the addition of acids, such as trichloroacetic acid or picric acid, due to the formation of insoluble salts, a property used in analytical techniques to separate proteins from solutions. The solubility of proteins can be reduced in a narrow pH range called the isoelectric point, when they are electrically neutral. When the temperature varies between 40°C and 50°C, the solubility of the enzymes increases. At temperatures above these, the tertiary structure is disrupted, and the protein is denatured and loses its activity [9].

6. Factors that inhibit enzyme activity

The stability of enzymes is a very important factor that must be taken into account when administering them in the course of the technological process. Enzymatic reactions are influenced by factors such as presence and concentration of enzymes and substrate, pH level, temperature, pressure, and presence of inhibitors and activators [2, 6, 9, 11]. The reaction rate varies in direct proportion to the concentration of the

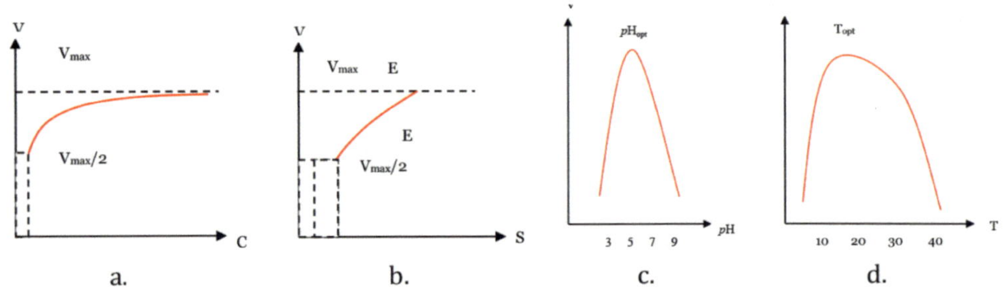

Figure 2.
Dependence of reaction rate on concentration—C (a), substrate—S (b), pH level (c), and temperature—T (d) [14].

enzyme and the substrate. Thus, by increasing the concentration of the enzyme above a certain threshold, the reaction rate will remain constant. On the other hand, by keeping the enzyme within constant limits while increasing the substrate concentration, the reaction rate will vary exponentially, as shown in **Figure 2**.

Consequently, the larger the available reactant surface, the higher the reaction rate, and as the particle size decreases, the total surface area increases. This allows for the participation of several reactant molecules in the chemical reaction. Most biological reactions occur in solution and their reaction rate is therefore directly proportional to the concentration of the reactant [5, 11]. **Figure 3** shows how the concentration varies depending on glucose isomerization and sucrose hydrolysis. The reaction rate is directly proportional to the concentration of the reactant under constant conditions of temperature and pH. In 1867, Guldberg and Waage posited a quantitative relationship between the molar concentration of the reactant (reactions) and the reaction rate [5].

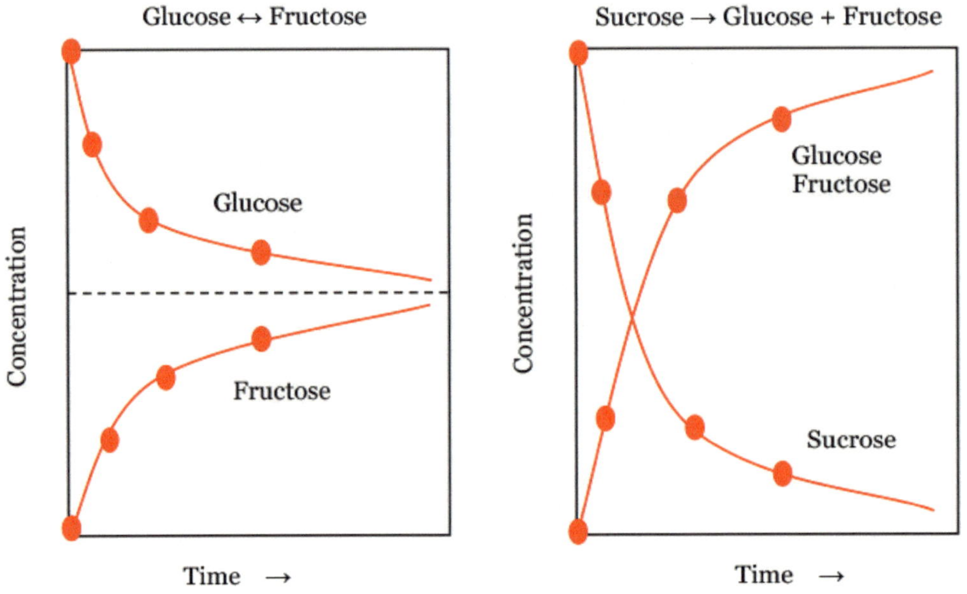

Figure 3.
Change in substrate concentration relative to reaction time. Fructose isomerization in fructose (left) and sucrose hydrolysis (right) [5].

The enzymatic reaction rate goes up with the rise in temperature to a maximum (optimal) threshold, only to go down with additional increases in temperature, which further cause denaturation of the enzyme [2, 5]. Thus, most enzymes show a maximum efficiency at temperatures between 35°C and 40°C for plant enzymes, and between 20°C and 30°C for those of animal origin. At temperatures above 60°C, enzymatic activity decreases or the protein component degrades completely. Enzymes are sensitive to the action of heat and change their properties due to temperature variation (they are thermolabile). For example, ribonuclease reduces its activity with increasing temperature but resumes it rapidly after cooling. However, some enzymes are more resistant and keep up their activity even at higher temperatures, as is the case of enzymes of various thermophilic bacteria, which remain active up to about 85°C [11]. There are also enzymes that can operate at very low temperatures, even in freezing conditions (e.g., β-galactosidase). Although the action of these enzymes is slower, it incurs lower costs both in terms of the amount of enzymes administered (smaller amounts of enzymes are required to meet the activation energy requirement) and energy consumption [6]. Endogenous enzymes of plants and fruits can be made up of isoenzymes with different thermal stabilities. The thermal stability of plant peroxidase isoenzymes has been long investigated to identify appropriate mechanisms and kinetic models for enzyme inactivation. For example, deamination of asparagine and glutamine residues, hydrolysis of peptide bonds to aspartic acid residues, oxidation of cysteine residues, thiol-disulfide exchange, destruction of disulfide bonds, and the chemical reaction between the enzyme and other compounds such as polyphenols can cause irreversible inactivation of enzymes at high temperature levels [11]. The dependence of the reaction rate on temperature can be expressed by means of the Arrhenius equation:

$$k = A_e^{-\frac{E_a}{RT}}$$

where k is a rate constant; A is the pre-exponential factor; E_a marks the activation energy; and R is the universal constant of the ideal gas at the absolute temperature T [2].

A high pressure has a major influence on enzymatic activity. Thus, at values over 3 kilobars, enzymes will be reversibly inactivated, while at a pressure of over 7 kilobars the process will be irreversible [11]. By applying high pressure, the activity of enzymes and the development of microorganisms are significantly inhibited, which allows for the protection of nutrients and flavor compounds. Microorganisms show extra sensitivity to high pressure, with their growth being inhibited at values between 300 and 600 MPa. On the other hand, a low pH level will emphasize this effect. However, bacterial spores can withstand pressures over 1200 MPa. In general, proteins are irreversibly denatured at ambient temperature by applying pressures above 300 MPa. Below this value, reversible changes in the structure of protein compounds occur. In the case of enzymes, even slight changes in the steric arrangement and mobility of the amino acid residues involved in catalysis can lead to their diminution and loss of activity [2]. The activity of enzymes is significantly influenced by the concentration of hydrogen ions in the reaction medium. Enzymes usually have a bell-shaped activity related to the pH profile (**Figure 2**). Decreased enzymatic activity on either side of the optimal pH can appear due to two causes. In the first case, the pH can influence the stability of the enzyme by inactivating it irreversibly. In the second situation, the pH can influence the kinetic parameters of the enzymatic reaction, namely: stability of the

ES complex, reaction rate, rate inhibition, or both. The pH dependence of enzyme-catalyzed reactions is similar to that of acid and base-catalyzed chemical reactions. In steady-state conditions, the integrated form of the Michaelis-Menten model is used:

$$K'_m \ln\frac{[S_0]}{S} + [S_0 - S = V_{max} t = k_{cat} \times [E_T] t$$

where K'_m is the apparent Michaelis constant; $[E_T]$ corresponds to the total enzyme concentration; $[S_0]$ and $[S]$ refer to the substrate concentration at time zero and time t, respectively; k_{cat} is the zero-order constant for the enzymatic reaction under conditions of substrate saturation; and t represents the reaction time [9].

The optimal pH value generates a maximum enzymatic activity and is influenced by the origin of the enzyme, type, and reaction medium. For example, pepsin has an optimal pH level between 1.4 and 2.5, while pancreatic amylase will show a maximum activity at pH 6.8. The relationship between enzymatic activity and pH level depends on the acid-base behavior of the substrate and the enzyme. Furthermore, the optimum pH for an enzymatic reaction is not the same as that of its normal intracellular environment. The activity of enzymes can also be inhibited by various chemical compounds, either endogenous (various metabolites) or exogenous (toxic agents, drugs). Ions and metal compounds, which are active as prosthetic groups or which ensure the stabilization of the configuration of the enzyme or the enzyme-substrate complex, are the activators of enzymatic reactions [2]. Enzyme inhibitors are low molecular weight chemical compounds that have the ability to completely reduce or inhibit the catalytic activity of the enzyme reversibly or permanently. That is to say, an enzyme inhibitor is a substance that slows down an enzyme-catalyzed reaction. It can alter one or more amino acids required in the enzymatic catalytic activity. Most natural inhibitors react reversibly with the enzyme and are classified into two types: specific and non-specific. The most common enzyme inhibitors with a wide range of applications in the food industry include protease inhibitors, polyphenol oxidase, and amylase or lipase inhibitors. For instance, protease inhibitors are substances that act directly on proteases to lower catalytic velocity. They usually mimic the protein substrate by binding to the active site of the enzyme and are specific for the active site of a class of proteinase. Protease inhibitors are usually classified according to the class of protease they inhibit (cysteine, serine, aspartic, and metalloprotease inhibitors). Most extracellular protein inhibitors produced by microorganisms belong to the genus *Streptomyces*. A number of pathogenic gram-negative bacteria, such as *Escherichia coli, Klebsiella pneumoniae, Serratia marcescens*, or *Erwinia chrysanthemi*, appear to be able to get protection against their own proteases by producing periplasmic protease inhibitors, such as ecotin [2]. Succinate dehydrogenase, which is responsible for the catalytic reactions involving the transformation of succinic acid into fumaric acid, is inhibited by dicarboxylic acids: malonic, malic, and oxalic [11]. In addition, food may be contaminated with pesticides, metal ions, and other chemicals in a polluted environment that may become inhibitory in certain circumstances [2].

7. Use of commercial enzymes in food and non-food industry

People have used enzymatic systems since ancient times, albeit with scarce information about them, to preserve food or ferment food or bread [1, 15]. Numerous

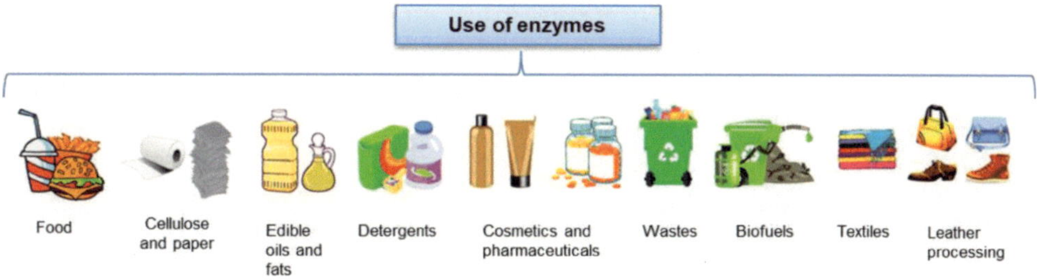

Figure 4.
Use of enzymes at industrial scale.

desired or undesirable changes in the aromatic profile and physicochemical properties of untreated fruits, vegetables, oilseeds, cereals, and food of animal origin are catalyzed by one or more enzymes. Whether activated intentionally or not, these enzymes influence the final quality of the food or drink in which they are present. Over time, major advances have been made in the field of enzyme chemistry with a focus on achieving a well-defined end product [3]. With technological progress, new enzymes have been developed that are characterized by a wide applicability and specificity [15]. Their use in industrial processes has shown increasing promise as they can eliminate the need for high temperatures, extreme pH values, and organic solvents and, at the same time, ensure high substrate specificity, low toxicity, high purity of the final product, low environmental impact, and easy inhibition of enzymatic activity [15]. **Figure 4** indicates some industrial applications of enzymes in food and non-food fields.

In the food industry, this technology allows for diversifying assortments and obtaining new products, improving nutritional value, reducing production costs, optimizing processing, and reducing the amount of waste, plus new solutions for food and packaging safety [6]. Enzymes are commonly used in the fruit processing industry to improve the pressing yield, extract and improve color and flavor characteristics, and clarify and decompose insoluble carbohydrates (pectins, hemicelluloses, and starch).

Enzymes play a key role in the production of beer and whiskey by helping to extract the sugars needed for fermentation, viscosity control, and to increase stability under storage conditions. Moreover, in the technology of beer, the administration of various enzymatic preparations can lead to a dietary product with low-calorie intake. Enzymes contribute significantly to improving the quality and stability of wines, reducing the period of alcoholic fermentation, promoting the clarification process, and ultimately facilitating filtration. The food industry is currently experiencing a growing trend in the demand for high-quality foods and beverages with outstanding, healthy sensory characteristics at competitive costs. The global market for enzymatic preparations used in the food industry, including beverages, reached approximately $1.69 billion in 2018, with growth expected to continue in the coming years, which poses a challenge to producers aiming to obtain innovative products. Most existing biotechnological applications are of microbial origin. Microbial enzymes are superior to those from animal and plant sources due to ease of production and genetic manipulation, various catalytic activities, etc. [8, 15]. The microorganisms used to produce the enzymes include about 50 bacteria considered safe by the FDA (GRAS - generally recognized as safe) and also fungi. The bacteria are mainly *Bacillus subtilis, Bacillus licheniformis,* and various species of *Streptomyces*. Fungi are usually of the genus *Aspergillus*, *Mucor,* and *Rhizopus*. Microorganisms can be grown in large quantities in a relatively short period of time by means of well-established fermentation methods. Large-scale production of microbial enzymes has many economic advantages due to cheap culture media and short fermentation stages [8, 9]. Globally, enzymes such as α-amylase, glucoamylase, lipase, pectinase, chymosin, and protease are most commonly used in the food processing industry. α-Amylase contributes to the transformation of starch into dextrins and is used in the production of corn syrup for various applications, such as sweetening various foods. In the production of high-quality beer, glucoamylase (hydrolytic enzymes) transforms dextrins into glucose by converting residual dextrins into fermentable sugars [6]. Proteases are of particular interest in the food industry due to their specific properties, such as high production yield, substrate specificity, high activity, and environmentally friendly nature. Also, the activity of these enzymes occurs in a wide range of temperatures (20 °C–80°C) and pH values (pH = 3–13), which increases its scope [15]. Also known as proteolytic enzymes, proteases are the largest class of such compounds in the human genome. They have the property of selectively catalyzing the hydrolysis of peptide bonds. Proteases are available in a wide variety of microorganisms, plants, and animals. Microbial production offers many benefits in terms of technical and economic properties, such as higher yields in a shorter time and reduced costs, plus a higher overall productivity [15]. The main field of application of proteases is the dairy industry, especially cheese manufacturing. Renin was initially preferred in cheese manufacturing due to its high specificity, while microbial proteases produced by GRAS microorganisms, such as *Mucor miehei, Mucor pusilis, Bacillus subtilis,* and *Endothia parasitica,* appeared not long after. For many years, proteases have also been used to produce low-allergenic milk proteins as ingredients for baby milk formulas. Proteases can also be used for the synthesis of peptides in organic solvents. The food industry uses the invertase produced by *Kluyveromyces fragilis, Saccharomyces cerevisiae,* and *Saccharomyces carlsbergensis* to make candy and jam. β-Galactosidase (lactase), produced by *Kluyveromyces lactis, Kluyveromyces fragilis,* or *Candida pseudotropicalis*, is used to hydrolyze lactose in milk or whey, while α-galactosidase secreted by *Saccharomyces carlsbergensis* is used to crystallize beet sugar [8]. Aspartic proteases, which play a role in the degradation of protein materials, comprise a small group of enzymes, among which cathepsin, renin, and pepsin are predominant. Their applications are well established in food processing in the manufacture of both traditional and modern products and are now being extended to new fields. They are widely used in cheese making, wine preservation, and also for clarifying beverages [15].

Cysteine protease, also known as bromelain, is isolated from the stem, fruit, or other parts of pineapple plants. It has a wide range of uses, from industrial to pharmaceutical domains. For most industrial applications, conventional production methods, such as extraction, concentration, and drying, are used. However, state-of-the-art applications in the pharmaceutical industry involve a much higher purity of bromelain, which is obtained through chromatographic methods, such as gene filtering or affinity chromatography [15].

Asparaginases are among the most widely clinically used enzymes, particularly in treating various cancers, insofar as they convert asparagine into aspartic acid and ammonia. Similarly, there has been a steady interest in their capacity to minimize the content of acrylamide in foods containing starch, and fried or baked products. Acrylamide is generated as a by-product of Maillard reactions between asparagine and reductive sugars. Reactions usually occur at temperatures above 100°C, being intended to alter the chromatic and aromatic profile of starchy foods, whether fried or baked. In 1994, acrylamide was first classified as group B2 - a possible carcinogen by the International Agency for Research on Cancer. Extensive efforts have been made to reduce the formation of acrylamide during baking or frying by incorporating the asparaginase. When used to reduce the formation of acrylamide in food, asparaginase can be isolated from fungal species and is considered safe, as it presents high specificity and minimal activity compared to glutamine. The main disadvantage of using asparaginase comes from marketing restrictions in some countries due to the associated problems at the industrial level. The incorporation of asparaginase in the food industry requires extensive research on the enzymatic effect and pre-/post-processing conditions. Purification of the enzyme needs extensive attention, as it influences the attenuation activity of acrylamide [15].

Lipases are universal enzymes that are present in all living things (plants, animals, fungi, and bacteria). Their basic function is to catalyze the hydrolysis of lipids into free fatty acids and glycerol at the interface of aqueous and organic solvents. Lipases catalyze a wide range of reactions that are significant from an industrial point of view, and present enantio-selectivity due to which they come to be seen as indispensable in food, pharmaceuticals, biofuels, detergents, cosmetics, leather industry, biosensor production, etc. [15]. For the production of fungal lipases, hosts such as *Aspergillus oryzae, Rhizomucor miehi*, *Thermomyces lanuginosus,* and *Fusarium oxysporum* [8] are used. In the food industry, lipases are used to improve the aromatic profile, reduce the time required for the maturation of cheeses, and obtain special products with superior qualities [6].

Cellulose, hemicellulose, pectin, and lignin are major components of the plant's cell wall. Hemicellulose is the second most abundant carbohydrate polymer on earth. α-L-arabinofuranosidase has a potential application in agro-industrial processes due to its synergistic effect with other hemicellulases. For example, α-L-arabinofuranoses are used in various industries: as a natural quality enhancer in bread manufacture; in the beverage industry to improve the aromatic profile of wines or to clarify fruit juices; in the production of pharmaceuticals, etc. [15].

Glucose oxidases are often used to remove oxygen from food or glucose from drinks for diabetics. These enzymes play an important role in defining the color, texture, flavor, and preservation of food. Lipases are used in the food industry to hydrolyze fats, improve taste characteristics, reduce the feeling of bitterness, or enhance preservation [6]. Lacases are increasingly used in various industrial oxidative processes, such as delignification, bioremediation, modification of plant fibers, ethanol production, biosensors, biofuels, etc. Industrial uses involve an increase in enzyme immobilization, usually from a heterologous host, such as *Aspergillus* spp. [8].

Enzymes are also used in a wide range of agro-biotechnological processes, and their main use is in the production of supplements to improve the nutritional quality of animal feed. For example, the use of phytases in agriculture as an ingredient in animal feed aims to improve the absorption of phosphorus from plants during the digestion of monogastric animals. Thus, phytase enables the release of phosphorus from plant matter, which contains about 2/3 of phosphorus as phytate, and reduces the phosphate load that impacts on the environment [8]. Another perspective on the use of phytases refers to human nutrition. It is known that ingestion of high amounts of food phytate severely hinders the absorption of important trace elements, such as iron and zinc, in the digestive tract. Due to this anti-nutritional effect of phytate, a large part of the population shows deficiencies with regard to these nutrients. There are two ways to reduce phytate dietary intake and its negative effects. One is to develop low phytate cultures by disrupting inositol polyphosphate kinases or other mutations in phytic acid biosynthesis. Although this approach has validated its main objective, low phytate maize and soybeans have been shown to have diminished seed yield and germination. Supplementing phytases in foods for human consumption is a more effective way to reduce the negative effect of phytate. To this end, Fujita *et al.* [16] tested a mutant strain of *Aspergillus oryzae* with high phytase activity in beer production. Haros *et al.* [17] used exogenous microbial phytase as an additive in bread making to improve physical and baking parameters, such as dosing time (24% reduction) width/height ratio of bread slices (5% reduction), specific volume (21% increase), and crumb firmness (28.3% reduction). While commercial use is under continuous testing, the potential role of thermophilic phytases as potent additives in the cellulose and paper industry has been suggested. Furthermore, phytase could act synergistically with xylanase in the preparation of multienzymes in xylanase-producing microorganisms such as *Streptomyces cupidosporus* [17].

In the chemical industry, the use of enzymes sometimes involves lower energy consumption, increased catalytic efficiency, much smaller amounts of waste and by-products, and lower volumes of wastewater. Hydrolases and ketoreductases that are stable in organic solvents are usually used for this purpose. They can also be used to produce various compounds, such as L-amino acids. About 150 biocatalysts are used in the chemical industry and are developed with the broadening application of genomic and protein engineering [8].

Enzymes are equally important in the pharmaceutical industry. For instance, penicillin acylases are used in the preparation of β-lactam antibiotics, such as semi-synthetic penicillins and cephalosporins. This group of antibiotics accounts for about 60-65% of the total antibiotic market. Enzymes are also involved in the preparation of chiral drugs and peptide synthesis. Furthermore, esterases, proteases, lipases, and ketoreductases are used in the preparation of chiral alcohols, carboxylic acids, amines, and epoxies [8].

8. Use of commercial enzymes in enology

The use of enzymatic preparations in winemaking is becoming more common in view of their many time-confirmed technological advantages. Endogenous enzymes play a key part in grape ripening/maturation [18]. They act by degrading the cell wall to favor the dissolution of vacuolar contents. The role of endogenous enzymes is incomplete since it is limited by winemaking conditions, such as pH of the must and insufficient activity due to the limited timespan of pre-fermentative treatments [19]. For these reasons, enzymatic preparations of an exogenous nature are often used in the technological process of wine depending on the winemaker's purpose. A sound knowledge of the nature and structure of macromolecules in must and wine offers new perspectives for the administration of enzymes in winemaking, especially in what concerns pressing, clarifying, filtering, and extracting various constituents with a role in defining the organoleptic characteristics of wine and its stabilization processes [18]. Moreover, the administration of such oenological products ensures the optimization of the process through a rigorous control on the quality of operations by allowing for superior loading of pressing and centrifugation equipment, reducing pressing times, favoring decantation and clarification of pressed juice, reducing energy consumption, and leading to an overall increase in production efficiency. The dosage of enzymes depends on the degree of ripeness of the grapes and the target one has in mind. For red wines, a larger dose variation is possible depending on incubation time. The enzymatic activities involved in the hydrolysis of pectic substances are

DOI: http://dx.doi.org/10.5772/intechopen.105474

carried out by pectin esterase, polygalacturonase, pectin lyase, rhamnogalacturonase, rhamnogalacturonan acetylesterase, arabinase, and galactanase. Other enzymatic activities come from hemicellulose and cellulose and are usually present in different amounts in the basic preparations of pectinases. The combined action of all these enzymes leads to partial hydrolysis and solubilization of neutral acid and polysaccharides in grapes [18]. Most enzymes are present in enzymatic preparations as isoenzymes and act differently depending on pH level, optimum temperature, and degree of pectin esterification. The most commonly used commercial enzymes are pectinases, glucanases, and glycosidases and less frequently lysozymes and urease [18].

Generally, the use of enzymatic preparations that have been purified to remove cinnamyl esterase activities is recommended for the production of white and rosé wines. Enzymatic preparations used in the manufacture of white wines are not thought to have such activity since it is already generated in nature by the species *Aspergillus niger* and *Botrytis cinerea,* and it is responsible for the hydrolysis of *p*-coumaric and ferulic acids, which, following decarboxylation, leads to the formation of 4-vinylphenol and 4-vinylguaiacol. These compounds are responsible for the medicinal odor in white wines specifically [18]. Lao *et al.* [20] presented that using purified pectolytic enzymes makes it possible to reduce the concentration of 4-vinylphenol in wines obtained from Sauvignon blanc grapes by more than 50%. Enzymatic preparations can be supplied in granular, liquid, or powder form. The latter has disadvantages due to the allergenic potential of enzymatic dust. The granular form has the double benefit of lacking preservatives and having good stability during storage, while liquid enzymes generally contain preservatives [18]. The concomitant use of bentonite and enzymatic preparations is to be avoided, as the enzyme will be inhibited due to the specific adsorption of bentonite and the latter will have reduced effectiveness due to blockage of active centers by the enzyme protein. Bentonite treatment should preferably take place after enzymatic treatment. Bentonite gel will help flocculate enzymatically hydrolyzed pectins. The activity and efficiency of an enzyme vary widely, depending on temperature and pH. Accordingly, pectinases can be administered at temperatures ranging between 10°C and 55°C. At temperatures below 10°C, the dose of the preparation should be increased, while at above 55°C the enzyme will be inactivated. β-Glucanases can only be used at temperatures above 15°C, as they require a longer incubation time. Enzyme dosage should also be increased with low pH values. Enzyme activity is not inhibited by optimal doses of sulfur dioxide in wine. With red wines, to the extent that inhibition of enzymatic activity may occur under the action of phenolic compounds, this allows for an increase in the dose of product administered. Alcoholic concentrations up to a level of 14% vol. do not impact negatively the action of enzymes. On the contrary, they may have an activating role on β-glucanases used to release flavor compounds [18]. The main benefits of using enzymes in the winemaking process are to do with their specificity of action, that is, less likely to produce unwanted secondary substances; their biodegradable nature and low impact on the environment; their capacity to get activated in conditions of low temperature, neutral pH, and normal atmospheric pressure; a significant reduction in energy consumption. Besides the many benefits, some unwanted activities of commercial preparations used in winemaking have been reported. They show high sensitivity to changes in physicochemical environmental conditions and can be distorted with relative ease (temperature, pH, infestations), which leads to an increase in the concentration of methanol during alcoholic fermentation under the action of methyl ethyl esterase. The action of cinnamyl esterase, present in enzymatic preparations based on pectinases, is responsible for the formation of a larger number of volatile compounds

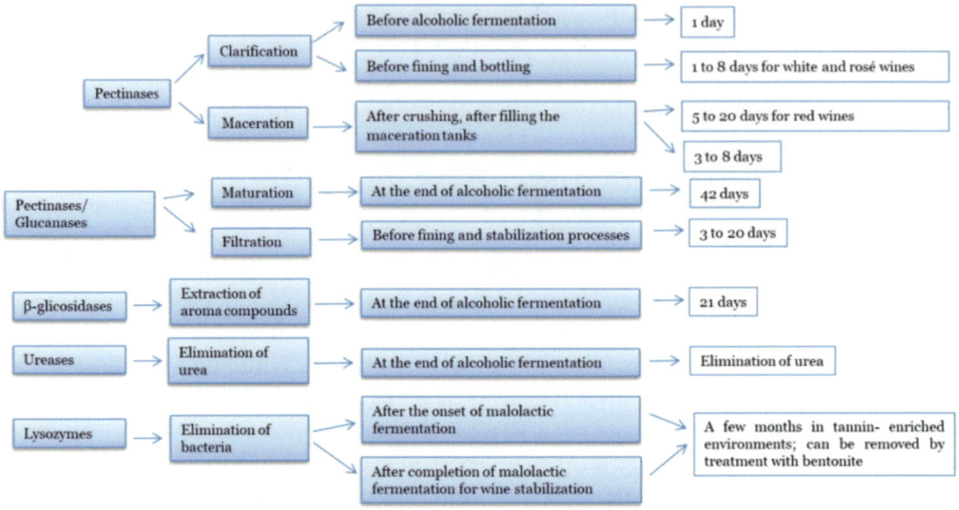

Figure 5.
Enzymes preparations usually used in enology.

[21]. These preparations are considered technological aids which are not found in the final product. **Figure 5** illustrates the main enzymatic preparations used in the technology of winemaking and some recommendations for their administration.

Comprehensive knowledge of regulations governing all treatments administered during the production stage is required, including timings, legally allowed amounts, and method of use. The use of enzymatic preparations in the beverage industry must comply with the regulations and recommendations of the International Organization of Vine and Wine, the Association of Manufacturers of Fermentation Enzyme Products, the World Health Organization, the United Nations Food and Agriculture (FAO), and the Food Chemical Codex [18].

8.1 Action of enzymes on the reaction yield obtained by pressing the must

The extraction yield of the juice can be significantly improved under the action of enzymes. Commercial preparations show various enzymatic activities at low pH values (pectin methyl esterases, polygalacturonases, pectin lyases, and hemicelluloses). These preparations may also contain various glycosides and proteases which are responsible for secondary transformations. Therefore, it is necessary to ensure a high degree of enzyme purity [19]. Pectinases are considered to be among the most important enzymes in the commercial sector, especially in the processing of fruit juices, that is, as adjuvants for the clarification and stabilization of juices, and to obtain a high yield as well. The degradation of cell walls under the action of pectinases allows for the wider diffusion of the constituents inside the vacuoles and facilitates a better extraction of the must during pressing [18]. The outcome depends on the amount of grape pectin, which varies according to the degree of maturation and grape variety, the enzymatic preparation administered (the type of enzymatic activity), and the conditions of administration (specific incubation time, pH of the environment, temperature, and presence of inhibitors). If pectinases are applied to grapes before pressing, they will increase the yield of juice extraction and color compounds [21]. The increase in pressing yield can reach at least 10%, in correlation with a reduction of up to 20–50% in the time needed for pressing, depending on the quality of grapes and the targeted result. When enzymes are applied without pre-fermentative maceration, their action mainly occurs at the time of pressing. At the maceration-fermentation stage, the enzymes are added immediately after the reception of the grapes. This improves the pressing efficiency and also the enzyme's degree of action. Pre-maceration is usually performed for about 3–4 hours at temperatures around 20°C, and 6–10 hours at temperatures below 15°C [18].

8.2 Influence of enzymes on wine clarification

During the processing of white and rosé wines, and especially after pressing, the must is rich in solid particles. Negatively charged pectin molecules form a protective layer around positively charged solid particles and keep them in suspension. Excessive turbidity of the must lends an herbaceous aroma to the wine, not to mention the hydrogen sulfide odor and a high content of isoamyl alcohol [18]. Consequently, clarification of the must before the alcoholic fermentation is particularly important as it considerably reduces the formation of aromatic compounds that give the wine unpleasant spicy notes or a salty sensation. Enzymatic preparations for clarification have predominantly pectolytic activities. Hydrolysis of pectic substances (**Figure 6**) leads to a significant reduction in the viscosity of the must [19]. During winemaking operations, segments of grape pectic compounds are released into the must after crushing and pressing. They form colloids that reduce or prevent the sedimentation of solid particles, especially skin fragments. Removal of solid particles is an important operation in the technology of obtaining white wines. Enzymatic hydrolysis of pectic structures is considered the most efficient method of decomposing colloids as it allows for the separation of captured solid particles. The presence of polygalacturonase and pectinase activity in grapes favors the clarification of the must after crushing. However, the activity of these enzymes is often insufficient since the time needed to obtain optimal clarification under the action of grape pectinases cannot compare with the time spent in classical winemaking processes. Therefore, for improving the efficiency of the clarification process, commercial preparations based on pectinases may be added [21]. Such preparations are added before the fermentation of musts of white varieties obtained after pressing and without prior maceration to accelerate clarification. Enzyme administration is recommended as a pre-fermentative

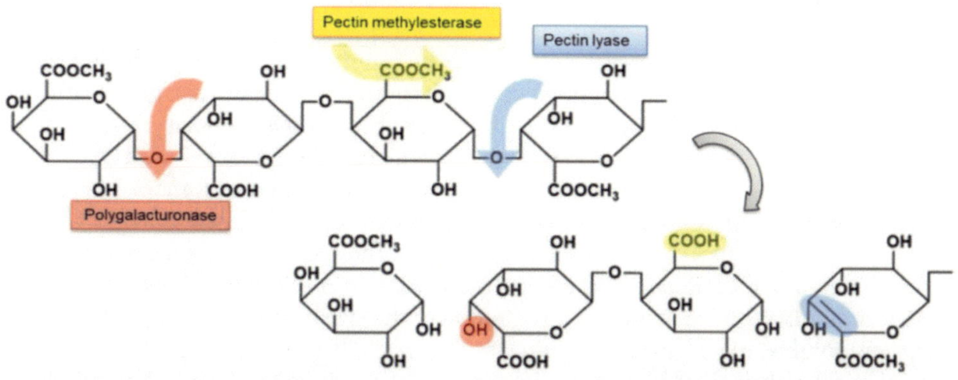

Figure 6.
Action of enzymes on grape pectin chains.

treatment because the high levels of alcohol resulting from fermentation tend to inhibit enzymatic activity.

Moreover, the use of pectolytic enzymes in wine technology is often associated with the maceration after heating the harvested grapes technique for red wines. This involves heating the must to 50°C and maintaining it at this temperature for several hours to solubilize the anthocyanins in the skin. The procedure sees the extraction of procyanidins in excess, which imprints astringency on the wine. In this way, the wines acquire an intense color but are not suitable for long-term aging. During heating, large amounts of pectin can be extracted from grapes, a phenomenon that does not occur in traditional processing. It becomes therefore necessary to administer a pectolytic preparation to reduce the viscosity of the must and remove the colloidal protective action of macromolecules with six carbon atoms (e.g., hexanol and hexanal) [19]. Following this process, the extraction of anthocyanins is intensified due to the decomposition of the cellular structure by the enzyme, which allows for easier dissolution of pigments. In traditional winemaking, the use of pectolytic enzymes triggers a significantly accelerated release of pigments, while maceration time can be shortened from 4 to 2 days. A potential disadvantage of this process is that the anthocyanins in the wines produced in this way can be unstable due to the hydrolysis of anthocyanin glycosides into their much more unstable aglycone forms. Secondary activities of enzymatic preparations are considered responsible for this glycosidic action [19]. The clarification of the must is carried out in three stages. The first stage is depectinization, characterized by the partial decomposition of pectins and the reduction of the must's viscosity. The second stage, flocculation, is described by an increase in turbidity and the formation of insoluble complexes. The third stage, sedimentation, is mainly characterized by a strong reduction in the turbidity and precipitation of complex molecules. Enzymes improve the first stage, thereby helping to accelerate the subsequent steps [18]. Significant improvements in clarification's degree have been reported with the use of pectolytic enzymes, β-glucanases, or proteases. Of these, proteases have been studied as an alternative to bentonite treatment, which would induce many chemical changes in the environment. Thus, Mojsov *et al.* [21] highlighted the degradation of enzymes that cause wine turbidity by administering enzymatic preparations based on lysozyme obtained from *Botrytis cinerea.*

8.3 Impact of enzymes on wine filtration

Enzymes for maturation and filtration consist mainly of pectinases and β-glucanases. Excess colloids are able to prevent filtration. Pectinases partially hydrolyze grape's polysaccharides and release smaller polysaccharide fragments. The latter usually presents a linear molecular structure; given the fact that these fragments can obstruct the different stages of wine filtration, their elimination before filtration is necessary. β-Glucanases hydrolyze glucan-type polysaccharides from *Botrytis cinerea* or yeast cell walls. Such polysaccharides are characterized by a high molecular weight and prevent or even make filtration impossible. The glucans released in wines by yeast (*Saccharomyces cerevisiae*) depend on the media used for yeast fermentation. At the same time, β-glucans can stimulate the extraction of certain macromolecules as mannoproteins which have an important role in stabilizing proteins in wines. A reduction in the size of these components makes them more soluble, maintains the colloidal structures in the wine during filtration, and diminishes the risk of filter blockage. The administration of enzymatic treatments can result in volumes up to five times higher during a filtration cycle, which helps to increase filtration efficiency with a reduction in costs, without affecting the sensory properties of the wine. These hypotheses were also confirmed by Mojsov *et al.* [21]. It is recommended that these enzymatic preparations be administered at the end of the alcoholic fermentation at a temperature above 15°C [18].

8.4 Action of enzymes on lactic acid bacteria

Due to their antibacterial action, lysozymes can inhibit the growth of bacteria, lactic acid bacteria in particular. Lysozyme administration can be an alternative to reducing the dose of SO_2 in low pH white wines [10]. These enzymes are able to degrade the cell walls of lactic acid bacteria - among other types, which make them an effective tool in preventing malolactic fermentation and increasing the stability of wines. The maximum regulated amount is 0.5 g/L in must or wine [22]. These oenological preparations are obtained by extraction from egg whites. For this reason, wines treated with lysozymes have to be labeled as "potentially allergenic." As pointed out in the literature, lysozymes reduce the concentration of biogenic amines in wine. In general, wines to which lysozymes have been administered are not to be conditioned by bentonite fining [10].

8.5 Influence of enzymes on color and basic physicochemical parameters of wines

The physicochemical properties of wines depend on the characteristics of the raw material, technological specificities, and the conditions in which fermentation takes place [10]. No significant influence has been reported on the main physicochemical parameters of wines [23, 24] following administration of enzymatic treatments. The visual characteristics of a wine depend on the degree to which its chemical structure and the compound's nature are able to absorb, transmit, and reflect light radiation from the visible spectral domain (between 380 and 750 nm). In recent years, oenological practices have considered enhancing the chromatic characteristics of wines by focusing on improving the extraction of color compounds. Although initially used to reduce turbidity and promote clarification, pectolytic enzymes have been demonstrated to be effective in intensifying color intensity and brightness, as well as the extraction of phenolic compounds [19, 21]. Similar results regarding the significant action of enzymes on the chromatic characteristics of white wines have been published by Ducasse *et al.* [25]. Guérin *et al.* [26] reported an improvement in

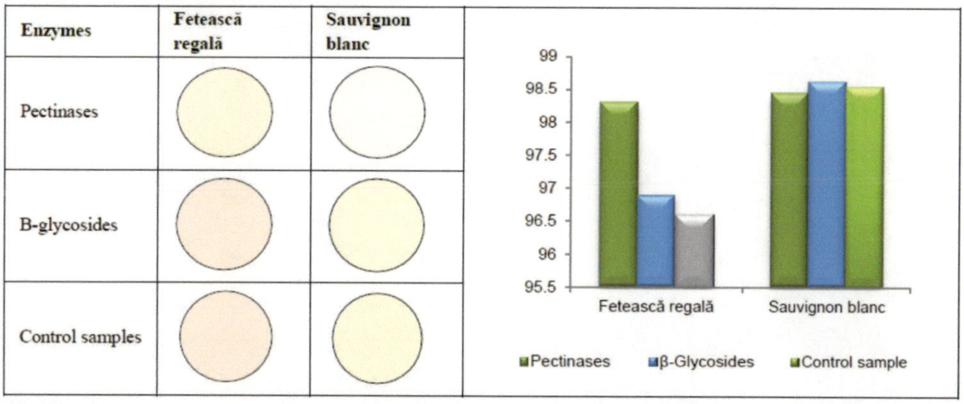

Figure 7.
Effect of enzymes on wine color and clarity [24].

wine brightness generated by the use of diverse enzymatic preparations. On the other hand, Bautista-Ortin *et al.* [27] obtained indecisive results in terms of changes in color parameters (intensity and hue), while Bozaran & Bozan [28] showed a reduction in color intensity and stability. These differences can be explained by the use of different enzymatic preparations and winemaking technology, but also by the presence of other uncontrolled factors in experimental studies. Along the same lines, the results published by Scutarașu [24] confirm the significant impact of using various enzymatic treatments on the values of the chromatic parameters of wines in the sense that a higher level of clarity is obtained (**Figure 7**). Bentonite treatment usually generates a significant decrease in the main chromatic parameters (clarity, chromaticity, and saturation) and increased values for tonality.

8.6 Impact of enzymes on wine phenolic compounds

The phenolic compounds in wine may originate in both grapes and external sources, such as the wood of the barrel in which they are stored and the cork used for bottling; alternately, they can appear after the administration of various oenological treatments. **Figure 8** represents the influence of enzymatic preparations on the content of phenolic compounds in some white wines studied by Scutarașu [24].

The level of these compounds depends on plant characteristics, analyzed variety, geographical location, specific year and harvesting procedure, and winemaking practices [10]. Phenolic compounds belonging to the group of flavones and flavonoids, especially hydroxycinnamic constituents (caffeic, *p*-coumaric, and ferulic acids) are mainly responsible for the color of white wines. In addition, the most common flavonoid derivatives in white wines are represented by quercetin, hesperidin, kaempferol, and rutin [29]. The proportions of the phenolic compounds are variables, participating in numerous physical, chemical, and biochemical processes. As a rule, in the first phase of the fermentation process, the oxidation of phenolic compounds that come from the raw material occurs under the action of enzymes. Some phenolic compounds may participate in the polymerization reaction with various flavor compounds. Hydroxycinnamic acids are involved in many oxidation reactions. Phenolic acids have proven to be important markers for Fetească regală and Sauvignon blanc varieties

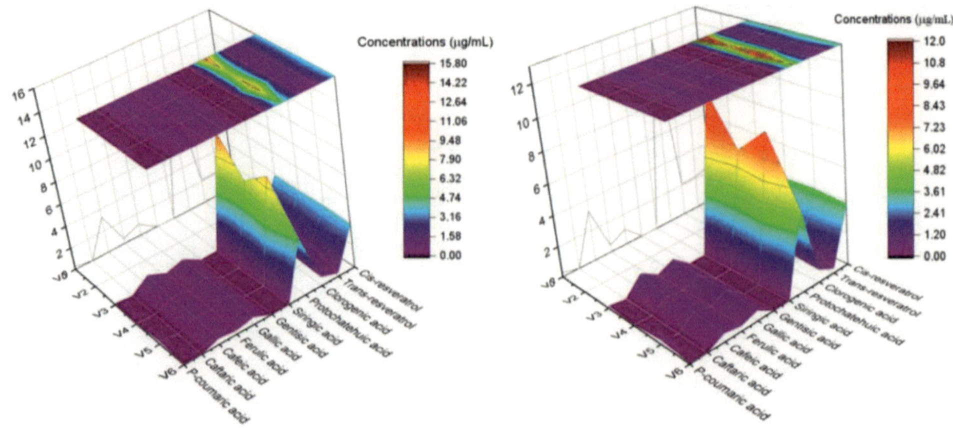

Figure 8.
Influence of enzymes on phenolic profile of Sauvignon blanc (left) and Fetească regală (right) [24].

from different wine regions of Romania and France [30]. The effects of enzymatic treatments on the chemical composition of wines have been studied intensively and far-reaching research on the influence of similar oenological products [27, 31, 32] reported significant increases in the phenolic content of wine. In general, the extraction of phenolic compounds occurs with the maceration of must and during alcoholic fermentation, and it depends on the variety and quality of the grapes and on winemaking technology. The effect of fungal laccase has been studied extensively due to its capacity of reacting with a wide range of phenolic compounds. Lacasse treatment is likely to increase the effect of conventional stabilization treatments [18]. Pectinases have been shown to be effective in enriching the medium in protocatechuic, caftaric, *trans*- and *cis*-resveratrol acids with the Sauvignon blanc variety, and in *p*-coumaric and gentisic acids with the Fetească regală variety (both from Iași vineyard, Romania) on condition they are administered in the must at the beginning of alcoholic fermentation [24]. Fining wines (previously treated with enzymes) with bentonite leads to lower values of phenolic compound concentrations. This phenomenon is due to the indirect adsorption effect of protein-binding phenolic compounds [10].

8.7 Effect of enzymatic treatments on wines' amino acids level

Wine amino acids can result from the degradation of grape proteins following the metabolism of yeasts and lactic acid bacteria, and from the autolysis of yeasts and bacteria. The profile and concentration of these compounds in wines can be influenced by several factors, such as grape variety, cultivation (treatments with nitrogen), and winemaking technology (e.g., maceration-fermentation process), as a result of amination and transamination of aldehydes and ketones, etc. Amino acids are particularly important for the formation and development of wine aromas (they are metabolic precursors of higher alcohols, volatile acids, and esters), and prove to be major factors in determining the authenticity and typicality of beverages. Insufficient amounts of such compounds can lead to incomplete fermentation and undesirable changes in the wine, such as hydrogen sulfide production and increased acetic acid proportions [33]. Amino acid concentration is also an important criterion for classifying wines according to their composition characteristics [34]. These compounds are highly reactive, being precursors of many flavor compounds, such as higher alcohols, esters, lactones, amines, etc. [10]. Most of the studies are focused on studying amino acids for classifying and differentiating wines according to variety, age, winemaking technologies, authentication, and typicity assessment [35]. According to the data presented by Cosme *et al.* [36], the synthesis of amino acids in grapes usually occurs at the end of the ripening stage, with proline and arginine being the main identified nitrogen compounds, followed by alanine, aspartic acid, and glutamic acid in smaller amounts. Numerous authors highlighted an important variation of the amino acid profile, depending on the grape variety and enzyme treatment. In this regard, Scutarașu [24] presented considerable amounts of some essential amino acids, such as histidine, isoleucine, phenylalanine, and tryptophan in wines treated with pectolytic enzymes preparations. The administration of pectolytic enzymes was more effective in the Fetească regală wines, in applied work conditions, although the β-glycosides generated the highest values of most amino acids in the Sauvignon blanc. Agustini *et al.* [37] obtained high proline and arginine concentrations in wine. The two compounds are not consumed during alcoholic fermentation due to anaerobic conditions and arginine metabolism. Beltran *et al.* [38] reported high amounts of asparagine (approximately 45 mg/L), lysine (16 mg/L) and

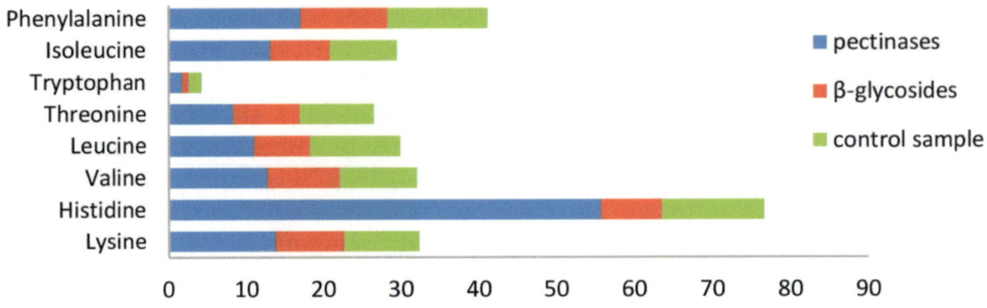

Figure 9.
Effect of enzymes on amino acid content in Fetească regală wines (mg/L) [24].

proline (approximately 500 mg/L). The data published by Scutarașu [24] indicate a major impact of both the type of enzyme administered and the grape variety on the characteristics of the wine (**Figure 9**).

Considerable amounts of some essential amino acids, such as histidine, isoleucine, phenylalanine, and tryptophan, were documented in the samples of Fetească regală and Sauvignon blanc (from Iași vineyard, Romania) treated with pectinases. As concerns the increased proportions of the amino acids under research, the administration of pectolytic enzymes was more efficient for Fetească regală wines, while β-glycosides generated the highest values of most amino acids in Sauvignon blanc samples when applied before alcoholic fermentation. Burin *et al.* [39] demonstrated a reduction in amino acid levels following the application of various fining and stabilization treatments, including the administration of pectolytic enzymes. Pinu *et al.* [40] have monitored the level of nitrogen compounds and their variation during the winemaking. Some amino acids, such as tyrosine, glycine, or arginine, were not exhausted by *Saccharomyces cerevisiae* during Sauvignon blanc alcoholic fermentation, which confirms previous observations on white wines made by Pinu *et al.* [40]. According to Cotea *et al.* [10], bentonite treatment can reduce wine protein levels by up to 15%.

8.8 Influence of enzymes on wines' volatile compounds

Wine's volatile compounds may originate from the grapes, being transferred to the must during processing, or may form during alcoholic fermentation, due to the biochemical reactions that occur in the wine. The administration of various pre-fermentative treatments significantly influences the aromatic profile of wines. The action of enzymatic treatments on the cell walls of grapes' skin is illustrated in **Figure 10**.

The free forms of varietal (terpenes) and combined (terpene glycoside) aromas are subject to oxidation and hydrolysis and are influenced by numerous biochemical and technological factors [10]. Most varietal aromas develop during fermentation, which suggests that the microbial species responsible for the fermentation process play a special role in releasing them from non-aromatic precursors. Enzyme preparations based on β-glycosidases can be added during winemaking to stimulate the extraction of volatile compounds from glycosidic bonds, especially monoterpenes, norisoprenoids, and benzenoids [21]. The varietal character of white wines is mainly defined by the presence of molecules with a characteristic odor, among which monoterpenic alcohols play a prominent role. These compounds are found in grapes as free, volatile, odorous molecules, and as non-volatile glycosidic precursors, known as bound terpenes. In many grape varieties, the number of bound terpenes may be

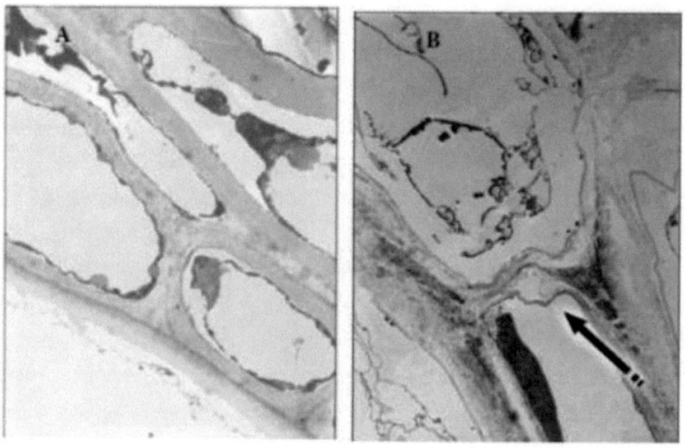

Figure 10.
The action of enzymes on the cell walls of grapes' skin—control (A) and after the administration of maceration enzymes (B) [18].

higher than the number of free terpenes. Consequently, the distinctiveness of wines could be increased by the release of terpenes with glycosidic bonds [21]. The presence of glycosylated precursors and volatile compounds in grapes was reported by Cordonnier & Bayonove [41]. In the late 1980s, enzymatic preparations containing glycosidases (β-glycosidase, α-arabinosidase, α-rhamnosidase) were developed to improve the aromatic profile of certain wines. These enzymatic preparations are usually added at the end of alcoholic fermentation and during wine transfer, in the absence of bentonite, to prevent inhibition of the enzyme. The optimum temperature for such enzyme treatments has to be in excess of 15°C, and an incubation time of a few weeks to a month is needed. The development of aromas has to be controlled by organoleptic analysis, and enzymatic action can be inhibited by the addition of bentonite. Small amounts of bentonite (20 g/hL) are usually sufficient to block the activity of enzymes completely [18]. Although much of a wine's aroma is attributed to alcohols and esters derived from yeast's metabolism, several grape varieties, such as Muscat, Gewürztraminer, Riesling, and Chardonnay, are characterized by specific, fragrant notes due to the presence of volatile monoterpenes such as linalool, geraniol, α-citronellol, and nerol [19]. These are released from the grapes during pressing, fermentation, and storage. Unlike many volatile fruit compounds, these compounds are glycosidically bound and are released slowly through acid hydrolysis exclusively, during wine aging. As the activity of endogenous glycosidases is very modest, there has been considerable interest in adding enzymes that promote the extraction of monoterpenes during winemaking. The secondary activities of fungal pectinases (e.g., *Aspergillus niger*) or extracellular glycosidases of various *Candida* yeasts may be used for this purpose [19]. Mateo & Stefano [42] pointed to the likely inhibition of β-glycosidase activity in the presence of ethanol and glucose. Enzymatic preparations have to be free of cinnamyl decarboxylase, which is instrumental in the formation of ethyl-phenols that give off an animal odor [19]. Numerous studies have indicated enrichment in the flavor profile of wines following administration of various enzymatic preparations. Thus, Masino *et al.* [31] obtained an increased level of the compound 4-vinylphenol in pectinase-treated samples. The action of pectolytic enzymatic preparations and β-glycosidases in obtaining white wines was also studied by Rusjan *et al.* [43] who recorded a significant increase in the concentrations of some monoterpenes (such as geraniol, nerol, linalool, or α-terpineol) compared to the control variant. Later on, Rusjan *et al.* [43] studied the effect of enzymatic preparations on white wines' terpenes. In this situation, the level of linalool did not increase significantly compared to the control sample. These results are supported by the use of enzymatic preparations with reduced α-rhamnosidase, α-arabinosidase, and β-glycosidase activity. Consequently, the choice of enzyme preparations suitable for the purpose proposed is of particular importance. Armada *et al.* [44] studied the effect of administering pectolytic enzymes in white wines obtained from the Albariño variety on the evolution of volatile compounds. All samples exhibited different aroma profiles, compared to the untreated ones, and samples obtained following the application of maceration enzymes showed the highest level for ethyl esters or phenethyl acetate. The use of maceration enzymes in combination with fining enzymes has been proved inappropriate due to the fact that glycosidic enzymes block the formation of flavor compounds. The main monitored components revealed differences between wines treated only with maceration enzymes (glycosidases) and wines to which other types of enzyme treatments were applied. Rocha *et al.* [45] reported a significant increase in the concentrations of geraniol, terpenoids, phenols, alcohols, and esters for the Maria Gomez variety, while no major changes in these compounds were observed for the Bical variety. The two varieties come from the same geographical area (Bairrada), which indicates that the extraction of flavor compounds under the influence of enzymes is closely related to the aromatic potential of the analyzed variety. According to other authors, the main volatile compounds of Sauvignon blanc wines are mercaptans (4-mercapto-4-methyl-2-pentanone), while others present methoxypyrazines (represented by 3-mercaptohexyl) as the defining compounds for the mentioned variety [46]. The aromatic profile of the wine depends on many factors, including the winemaking technology and the particulars of the geographical region. With reference to the study conducted by Scutarașu [24], the fining of wines (to which enzymes were added) with bentonite triggered changes in the proportions of volatile compounds depending on the compounds' class, grape variety, and administrated enzyme types. Regarding the level of carbonyl compounds in Sauvignon blanc wines, bentonite treatment led to increased quantities of acetoin (3-hydroxy-2-butanone) and benzaldehyde. Bentonite-treated Fetească regală samples exhibited reduced levels of acetoin. Other studies reported similar changes in these compounds [47]. Some authors have focused on the impact of bentonite on ethyl esters concentrations. For instance, Vincenzi *et al.* [47] reported a decreasing trend in the proportion of ethyl alcohol esters, being protein bound. Lambri *et al.* [48] reported a decrease in the content of ethyl butyrate and ethyl hexanoate in Chardonnay wines. Sanborn *et al.* [49] obtained a decrease in the level of ethyl decanoate and phenylethyl acetate in Gewürztraminer wines, while no changes were reported for Chardonnay wines. In the experimental samples obtained by Scutarașu [24], this hypothesis was confirmed by ethyl butanoate and ethyl dodecanoate in Sauvignon blanc samples and ethyl hexanoate, ethyl octanoate, ethyl 3-hydroxybutanoate, ethyl decanoate, and ethyl 4-hydroxybutanoate in most variants of Fetească regală wines, respectively. Moreover, regarding the level of fatty acids, the main precursors of aromatic esters, a decrease of butanoic, octanoic, and decanoic acids content in the Fetească regală samples was registered, correlated with an increase in the ratio of hexanoic and octadecanoic acid. In bentonite-treated Sauvignon blanc wines, 3-methylbutanoic, hexanoic, octanoic, and decanoic acids had higher values. The interaction of this treatment with fatty acids has been studied by several authors. Vincenzi *et al.* [47] also reported an increase in decanoic and dodecanoic acid concentrations as well as a decrease in the amount of octanoic acid in Muscat wines.

DOI: http://dx.doi.org/10.5772/intechopen.105474

Compounds	Tendency	References
Linalool	↗	[50]
Citronellol	↘	
α-Terpineol	↘	
Hotrienol	↗	
Geraniol	↗	[44]
Nerol	↗	
Propanol	↗	[50]
Isobutanol	↘	
I-Butanol	↘	[44]
Isoamyl acetate	↘	[50]
Ethyl hexanoate	↘	[50]
Furfural	↗	[44]
Ethyl lactate	↘	
Hexanoic acid	↗	[24]
Octanoic acid	↗	[24]

Table 1.
Effect of enzyme treatment on wine's volatile compounds.

McKinnon [33] showed a positive correlation between the level of decanoic acid and phenylalanine. **Table 1** presents the impact of enzymes on some volatile compounds in wines.

8.9 Effect of enzymatic treatments on sensory properties

As far as the consumer is concerned, the organoleptic characteristics of foods are the decisive factor influencing purchase choice. The administration of enzyme treatment is mainly aimed to enrich and improve the sensory profile. In direct correlation with the data presented in this chapter, major organoleptic differences have been reported between wines treated with various enzymatic preparations. Enrique *et al.* [51] indicated a significant increase in the intensity of the sensory descriptors studied in samples treated with pectolytic enzymes. Scutarașu [24] confirmed that pectinases can improve the sensory characteristics of wines compared to β-glycosides (**Figure 11**) and that the samples are generally characterized by the lowest intensity for some negative descriptors, such as phenolic, mineral, or bitter taste. This research highlighted that β-glycosides can give effective results when administrated before alcoholic fermentation, in must.

Sun *et al.* [52] obtained higher levels of acidic fruits, sweet fruits, and other notes in wine treated with enzymes. The application of *H. uvarum* extracellular enzyme enhanced fruity and floral aroma, especially the acidic fruits notes [52]. According to the data presented by Bautista-Ortin *et al.* [27], wines treated with pectinases had higher scores for their herbaceous, dryness, astringency, and bitterness characteristics, and showed lower equilibrium than the control sample. McKinnon [33] reported a positive correlation between leucine levels and fruity or floral notes in samples treated with pectolytic enzymes. Gonzáles-Lázaro *et al.* [53] indicated that pectolytic

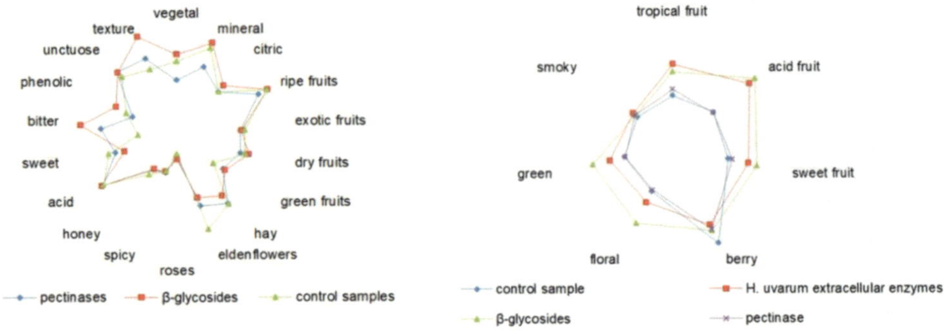

Figure 11.
Effect of enzymes on sensory properties of wines [24, 52].

enzymes did not show effective results in sparkling wines when these preparations are administrated on unripe grapes.

9. Conclusions

Wine quality is dependent on grape characteristics and winemaking technology. Enzymes' activity is influenced by their concentration, substrate, pH, temperature, pressure, and the presence of inhibitors and activators. Several authors confirmed the positive impact of using enzymes on wine quality. However, higher concentrations of phenolic compounds and amino acids and enriched volatile and sensory profiles can be obtained when enzyme preparation is used. Enzymes contribute to optimizing the technological process in view of improving the quality of the final product, while giving effective results when they are administrated at different moments in winemaking. Summing up all the above, enzymatic preparations will remain in focus for the near future to analyze possible new applications in the food and non-food industries.

Conflict of interest

The authors declare no conflict of interest.

Author details

Elena Cristina Scutarașu, Camelia Elena Luchian, Lucia Cintia Colibaba and Valeriu Cotea*
Iași University of Life Sciences, Iași, Romania

*Address all correspondence to: vvcotea@yahoo.com

References

[1] Aroca A, Rollenghem I, Schneider R. Enzymes application in white wine. In: Morata A, editor. White Wine Technology. USA: Academic Press; 2022

[2] Belitz H-D, Grosch W, Schieberle P. Food Chemistry. Berlin; New York: Springer; 2004

[3] Espejo F. Role of commercial enzymes in wine production: A critical review of recent research. Journal of Food Science and Technology. 2020;**58**(1):9-21

[4] Palmer T, Bonner P. Enzymes – Biochemical, Biotechnology and Clinical Chemistry. 2nd ed. UK: Woodhead Publishing; 2007

[5] Punekar NS. Enzymes: Catalysis Kinetics and Mechanisms. Singapore: Springer; 2018

[6] Kuddus M. Enzymes in Food Biotechnology, Production, Application, and Future Prospect. London: Academic Press; 2019

[7] Botella C, Webb C, Cantero D, Blandino A. Hydrolytic enzyme production by Aspergillus awamori on grape pomace. Biochemical Engineering Journal. 2005;**26**(2-3):100-106

[8] Sanchez S, Demain LA. Useful microbial enzymes – an introduction. In: Brahmachari G, Demain L, Jose A, Demain A, editors. Biotechnology of Microbial Enzymes. Academic Press; 2017

[9] Subin SR, Bharm SG. Enzymes: concepts, nomenclature, mechanism of action and kinetics, characteristics and sources of food-grade enzymes. In: Chandrasekaran M, editor. Enzymes in Food and Beverage Processing. USA: CRC Press; 2016

[10] Cotea DV, Zănoagă C, Cotea VV. Tratat de oenochimie. București: Romanian Academy Press; 2009

[11] Suzuki H. How Enzymes Work: From Structure to Function. Singapore: Jenny Stanford Publishing Pte Ltd; 2015

[12] Ogston AG. Interpretation of experiments on metabolic processes, using Isotopic Tracer Elements. Nature. 1948;**162**(4129):963-963. DOI: 10.1038/162963b0

[13] Poshyvailo-Strube L. Modelling and Simulation of Enzyme-catalyzed Reactions. Ucraine: National University of Kyiv-Mohyla Academy; 2015

[14] Nagodawithana TW, Reed G. Enzymes in Food Processing. San Diego: Academic Press; 1993

[15] Parameswaran B, Varjani S, Raveendran S. Green Bio-processes Enzymes in Industrial Food Processing. Singapore: Springer; 2019

[16] Fujita J, Fukuda H, Yamane Y, Kizaki Y, Shigeta S, Ono K, et al. Biotechnology Letters. 2001;**23**(11):867-871. DOI: 10.1023/A:1010599307395

[17] Haros M, Rosell CM, Benedito C. Use of fungal phytase to improve breadmaking performance of whole wheat bread. Journal of Agricultural and Food Chemistry. 2001;**49**(11):5450-5454. DOI: 10.1021/jf010642l

[18] Gómez-Plaza E, Romero-Cascales I, Bautista-Ortin AB. Use of Enzymes for Wine Production. Enzymes in Fruit and Vegetable Processing – Chemistry and Engineering Applications. USA: CRC Press; 2010

[19] Ribéreau-Gayon P, Dubourdieu D, Donèche B, Lonvaud-Funel A, Glories Y, Maujean A, et al. Handbook of Ecology. Hoboken, NJ: Wiley; 2020

[20] Lao C, Lopez-Tamames E, Lamuela-Raventos RM, Buxaderas S, Torre-Boronat MDC. Pectic enzyme treatment effects on quality of white grape musts and wines. Journal of Food Science. 1997;**62**(6):1142-1149. DOI: 10.1111/j.1365-2621.1997.tb12231.x

[21] Mojsov K, Ziberoski J, Bozinovic Z, Petreska M. Comparison of effects of three commercial pectolytic enzyme preparations in white winemaking. Applied Technologies and Innovations. 2011;**4**(1):34-38

[22] International Oenological Codex International Organisation of Vine and Wine [Internet]. Available from: https://www.oiv.int/public/medias/7790/codex-2021-en.pdf

[23] Samoticha J, Wojdyło A, Chmielewska J, Politowicz J, Szumny A. The effects of enzymatic pre-treatment and type of yeast on chemical properties of white wine. LWT - Food Science and Technology. 2017;**79**:445-453. DOI: 10.1016/j.lwt.2017.01.063

[24] Scutarașu E. Studies on Enzymes Impact on White Wines Technology from Iași Vineyard. Romania: Iasi University of Life Sciences; 2021

[25] Ducasse M-A, Canal-Llauberes R-M, de Lumley M, Williams P, Souquet J-M, Fulcrand H, et al. Effect of macerating enzyme treatment on the polyphenol and polysaccharide composition of red wines. Food Chemistry. 2010;**118**(2):369-376. DOI: 10.1016/j.foodchem.2009.04.130

[26] Guérin L, Shutter DH, Demois A, Chereau M, Trandafir G. Determination of activity profiles of the main commercial enzyme preparations used in winemaking. American Journal of Enology and Viticulture. 2010;**60**(3):322-331

[27] Bautista-Ortin AB, Martinez-Cutillas A, Ros-Garcia JM, Lopez-RocaJM,Gomez-PlazaE.Improving colour extraction and stability in red wines: the use of maceration enzymes and enological tannins. International Journal of Food Science and Technology. 2005;**40**(8):867-878. DOI: 10.1111/j.1365-2621.2005.01014.x

[28] Borazan AA, Bozan B. The influence of pectolytic enzyme addition and prefermentative mash heating during the winemaking process on the phenolic composition of Okuzgozu red wine. Food Chemistry. 2013;**138**(1):389-395. DOI: 10.1016/j.foodchem.2012.10.099

[29] Lengyel E, Silkolya L. Authenticity tests of hite wines from the Apold depression. Management of sustainable development. 2014;**6**(2):55-59. DOI: 10.1515/msd-2015-0007

[30] Merkytė V, Longo E, Windisch G, Boselli E. Phenolic compounds as markers of wine quality and authenticity. Foods. 2020;**9**(12):1785. DOI: 10.3390/foods9121785

[31] Masino F, Montevecchi G, Arfelli G, Antonelli A. Evaluation of the combined effects of enzymatic treatment and aging on lees on the aroma of wine from Bombino bianco grapes. Journal of Agricultural and Food Chemistry. 2008;**56**(20):9495-9501. DOI: 10.1021/jf8015893

[32] Scutarașu E-C, Luchian CE, Vlase L, Colibaba LC, Gheldiu AM, Cotea VV. Evolution of phenolic profile of white wines treated with enzymes. Food Chemistry. 2021;**340**:127910. DOI: 10.1016/j.foodchem.2020.127910

[33] McKinnon A. The impact of amino acids on growth performance and major volatile compound formation by industrial wine yeast [Internet]. 2013. Available from: http://scholar.sun.ac.za. 2013

[34] Pereira V, Pereira AC, Pérez Trujillo JP, Cacho J, Marques JC. Amino acids and biogenic amines evolution during the estufagem of fortified wines. Journal of Chemistry. 2015;**2015**:1-9. DOI: 10.1155/2015/494285

[35] Soufleros EH, Bouloumpasi E, Tsarchopoulos C, Biliaderis CG. Primary amino acid profiles of Greek white wines and their use in classification according to variety, origin and vintage. Food Chemistry. 2003;**80**(2):261-273. DOI: 10.1016/s0308-8146(02)00271-6

[36] Cosme F, Andrea-Silva J, Filipe-RibeiroL,MoreiraASP,MalheiroAC, Coimbra MA, et al. The origin of pinking phenomena in white wines: An update. BIO Web of Conferences. 2019;10.1051/bioconf/20191202013

[37] Agustini BC, Lima DBD, Bonfim TMB. Composition of amino acids and bioactive amines in common wines of Brazil. Acta Scientiarum Health Science. 2014;**36**(2):225. DOI: 10.4025/actascihealthsci.v36i2.20187

[38] Beltran G, Novo M, Rozes N, Mas A, Guillamon J. Nitrogen catabolite repression in during wine fermentations. FEMS Yeast Research. 2004;**4**(6):625-632. DOI: 10.1016/j.femsyr.2003.12.004

[39] Burin VM, Caliari V, Bordignon-Luiz MT. Nitrogen compounds in must and volatile profile of white wine: Influence of clarification process before alcoholic fermentation. Food Chemistry. 2016;**202**:417-425. DOI: 10.1016/j.foodchem.2016.01.096

[40] Pinu FR, Edwards PJB, Gardner RC, Villas-Boas SG. Nitrogen and carbon assimilation by Saccharomyces cerevisiae during Sauvignon blanc juice fermentation. FEMS Yeast Research. 2014;**14**(8):1206-1222. DOI: 10.1111/1567-1364.12222

[41] Cordonnier R, Bayonove C. Mise en évidence dans la baie de raisin, var. Muscat d' Alexandrie, de monoterpènes liés, révélables par une ou plusiers enzymes du fruit. Compte Rendus Academie des Sciences Paris. 1974;**278**:3387-3390

[42] Mateo JJ, Di Stefano R. Description of the β-glucosidase activity of wine yeasts. Food Microbiology. 1997;**14**(6):583-591. DOI: 10.1006/fmic.1997.0122

[43] Rusjan D, Strlič MS, Košmerl T, Prosen H. The response of monoterpenes to different enzyme preparations in Gewürztraminer (*Vitis vinifera* L.) wines. South African Journal of Enology & Viticulture. 2016;**30**:1

[44] Armada L, Fernández E, Falqué E. Influence of several enzymatic treatments on aromatic composition of white wines. LWT - Food Science and Technology. 2010;**43**(10):1517-1525. DOI: 10.1016/j.lwt.2010.06.009

[45] Rocha SM, Coutinho P, Delgadillo I, Cardoso AD, Coimbra MA. Effect of enzymatic aroma release on the volatile compounds of white wines presenting different aroma potentials. Journal of the Science of Food and Agriculture. 2004;**85**(2):199-205. DOI: 10.1002/jsfa.1937

[46] Roland A, Cavelier F, Scheinder R. Les thiols varietaux dans les vins: Point sur les voies de biogenese et incidence des itineraires de production et d'elaboration. Colloques Internationaux sur les aromes du vin (Project Vinaromas). Toulouse et Saragosse. 2012

[47] Vincenzi S, Panighel A, Gazzola D, Flamini R, Curioni A. Study of combined effect of proteins and bentonite fining on the wine aroma loss. Journal of Agricultural and Food Chemistry. 2015;**63**(8):2314-2320. DOI: 10.1021/jf505657h

[48] Lambri M, Dordoni R, Silva A, de Faveri DM. Effect of bentonite fining on odor-active compounds in two different white wine styles. American Journal of Enology and Viticulture. 2010;**61**(2):225-233

[49] Sanborn M, Edwards CG, Ross CF. Impact of fining on chemical and sensory properties of Washington state Chardonnay and Gewurztraminer wines. American Journal of Enology and Viticulture. 2010;**61**(1):31-41

[50] Ahumada K, Martínez-Gil A, Moreno-Simunovic Y, Illanes A, Wilson L. Aroma release in wine using co-immobilized enzyme aggregates. Molecules. 2016;**21**(11):1485

[51] Enrique M, Ibánez A, Marcos J, Yuste M, Martínez M, Vallés S, et al. β-Glucanases as a tool for the control of wine spoilage yeasts. Journal of Food Science. 2010;**75**(1):M41-M45. DOI: 10.1111/j.1750-3841.2009.01448.x

[52] Sun W-X, Hu K, Zhang J-X, Zhu X-L, Tao Y-S. Aroma modulation of Cabernet Gernischt dry red wine by optimal enzyme treatment strategy in winemaking. Food Chemistry. 2018;**245**:1248-1256

[53] González-Lázaro M, Martínez-Lapuente L, Guadalupe Z, Ayestaran B, Bueno-Herrera M, López de la Cuesta P, et al. Evaluation of grape ripeness, carbonic maceration and pectolytic enzymes to improve the chemical and sensory quality of red sparkling wines. Journal of the Science of Food and Agriculture. 2020;**100**(6):2618-2629

Chapter 5

Low-Alcohol and Nonalcoholic Wines: Production Methods, Compositional Changes, and Aroma Improvement

Teng-Zhen Ma, Faisal Eudes Sam and Bo Zhang

Abstract

Nonalcoholic wine (NW) has attracted the interest of winemakers and researchers in recent years, mainly due to the increasing market share of NW (≤ 1% alcohol by volume), the health risks associated with the consumption of wine, the global trend toward healthier lifestyles, and the uncompromising cardioprotective effects of NW. NW can be produced using several methods, particularly, dealcoholization of wines, which is mainly achieved by physical dealcoholization methods. However, the dealcoholization of wine has two major drawbacks. The first drawback is legal since the laws vary according to each country. The second disadvantage is technical since it is difficult to dealcoholize a wine while maintaining its original organoleptic characteristics. Both the aromatic qualities (volatile composition) and taste (sensory characteristics) of the dealcoholized wine (DW) tend to worsen the greater the decrease in its alcoholic strength. This makes the resulting wine have a different flavor and aroma. Improvement of the aroma of DW after dealcoholization could help wine producers limit undesirable effects and increase consumer acceptance. This chapter is focused on the popular techniques used in wine dealcoholization, their impact on the phenolic composition, volatile composition, sensory characteristics, and the state-of-the-art methods of improving the aroma profile of DW.

Keywords: wine, nonalcoholic wine, dealcoholization, volatile compounds, phenolic compounds, sensory quality, aroma profile

1. Introduction

Wine is an alcoholic beverage commonly produced from fermented grape juice. It can be categorized as dry wine, semidry wine, semisweet wine, sweet wine, still wine, and sparkling wine based on its sugar or carbon dioxide content. Depending on the production methods or materials used, it can also be classified as a special wine (including liqueur wine, carbonated wine, icewine, noble rot wine, floral wine, flavored wine, low alcohol wine, nonalcohol wine, and *V.amurensis* wine) [1].

In addition to the categorization of wine regarding its alcohol content, one with an alcohol content above 10.5% v/v, from 5.5 to 10.5% v/v, from 1.2 to 5.5%, from 0.5 to 1.2% v/v, or below 0.5% v/v may be classified as an alcoholic wine, lower-alcohol wine, reduced-alcohol wine, low-alcohol wine, or nonalcoholic wine (NW) [2, 3]. However, these classifications differ from one winemaking region or country to another [4]. Over the years, with health risk awareness and social demands related to road safety, consumer preferences are now shifting toward new product offerings and alternatives, with an increasing percentage of the adult population seeking lower alcohol wines more frequently. This has boosted the production and sales of nonalcoholic wines with the global nonalcoholic wine market worth over $10 billion and is still estimated to increase at a significant CAGR above 7% between 2019 and 2027, attaining a profit share of over $30 billion [5].

Lower, reduced, low, and NW wines can be produced at the various stages of wine production (pre-fermentation, fermentation, and post-fermentation stages) using several methods such as reduction of the juice fermentable sugars before fermentation, the reduction of alcohol production during fermentation, and the separation by membranes and thermal treatment after complete fermentation of the wine [6–9]. The latter methods, also known as physical dealcoholization (ethanol removal) methods are usually used after complete fermentation of the wine (i.e., at wine post-fermentation stage) and can achieve good results when used on a finished wine. Studies have reported the ability of some physical dealcoholization methods in preserving the phenolic compounds [10], volatile compounds [11], and sensory quality [12, 13] of the final wine product at certain levels of ethanol removal with a taste almost similar to the original wines (in the case of partially dealcoholized wines), contrarily to the former methods (commonly used before and during fermentation), which produces unbalanced wine products (high acidity, unfermented juice, and low fermentative aroma compounds) with legality issues in the case of the juice fermentable sugars dilution with water [14]. Wine produced by physical dealcoholization methods (i.e., alcohol removal from finished wines) is termed dealcoholized wine (DW). The dealcoholization of wine can be complete or partial. A completely DW is a beverage obtained exclusively from wine by dealcoholization with a final alcohol content below 0.5% v/v (resolution OIV-ECO 432-2012), while a partially DW is a beverage obtained exclusively from wine by dealcoholization with a final alcohol content ≥0.5% v/v (resolution OIV-ECO 433-2012). NW (< 0.5% v/v ethanol) produced from a finished wine by dealcoholization may be termed DW, whereas low, reduced, and lower alcohol wines (0.5–10.5% v/v ethanol) may be termed partially dealcoholized wines. In this chapter, we focus on the methods used for producing of NW from high-strength alcoholic wines after complete fermentation (wine post-fermentation stage), specifically, their impact on the aroma profile and sensory characteristics of NW. In addition, the state-of-the-art methods of improving the aroma profile of DW/NW are discussed.

2. Methods of lower, reduced, low, and NW wines production

The production of NW can be achieved by several methods as shown in **Figure 1**. These methods can be broadly classified into three groups based on the principle or mechanism of ethanol reduction and removal at the various stages of wine production, including reduction of fermentable sugars (pre-fermentation stage), reduction or limitation of ethanol production (fermentation stage), and ethanol removal by membrane separation or thermal treatment (post-fermentation stage) [6–9].

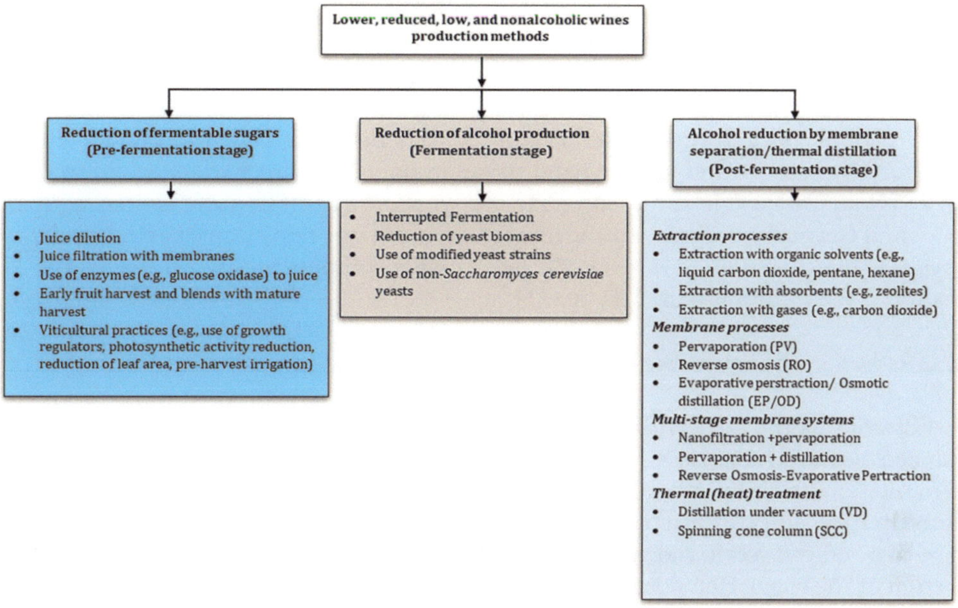

Figure 1.
Methods of lower, reduced, low, and NW wines production.

2.1 Decrease of ethanol production through reduction of fermentable sugar content

The reduction of fermentable sugars in the pre-fermentation stage of wine is one of the common methods for the production of wines with lower or reduced alcohol content. It includes techniques such as juice dilution [15, 16], juice filtration with membranes [17, 18], the use of enzymes (e.g., glucose oxidase) [19, 20], early harvest and blending with mature harvest [21], viticultural practices (e.g., use of growth regulators, reduction of photosynthetic activity, reduction of leaf area, preharvest irrigation) [22, 23].

2.1.1 Reduction of fermentable sugars by juice dilution

Juice dilution involves adding water to grape juice or mixing the juice with green harvest to reduce the concentration of fermentable sugars. In countries such as South Africa, New Zealand, Australia, and the United States of America (excluding California where it is only permitted for preventing stuck fermentations), water is only allowed as a processing aid. The substitution of grape juice with water or the direct addition of water to reduce the concentration of fermentable sugars has been effective in reducing the ethanol content of the final wine product by 4-6% v/v [15, 16]. Regarding the use of green harvest in juice dilution, this range of ethanol reduction is determined by the harvest date. The pre-fermentative substitution of a matured Shiraz juice (obtained from Shiraz grapes harvested at 25.5 °Brix) with water or direct water addition at rates of 10.2, 34.0, and 47.2% v/v resulted in lower alcohol wines with 14.5%/14.4%, 12.0%/11.7%, and 10.6%/9.6% ABV, respectively, after fermentation to dryness (<1 g/L of total sugar). The lower alcohol wines (10.6%/9.6% ABV) produced by substituting or diluting the juice with 47.2% v/v water decreased the total phenolics, anthocyanin concentration, tannin concentration, color density, and SO_2-resistant pigments compared with the control (15.5% ABV) [16]. Furthermore, sensory attributes such as "body," "astringency," "flavor intensity," and "alcohol" flavor were lower in wines with 9.6% ABV compared with the control, which was attributed to the alcohol concentration difference of 1% ABV between the lower alcohol wines produced by substitution (10.6%) and those produced by direct water addition (9.6% ABV) [16]. However, in some European winemaking countries such as France, Spain, Italy, and Germany, this practice is illegal because it can significantly affect most physicochemical parameters, phenolic and volatile components, and the sensory quality of the resulting wine [15, 16, 21].

2.1.2 Reduction of fermentable sugars by filtration of juice

Filtration of the juice with membranes is another method of producing lower or reduced alcohol wines, based on the principle of sugar reduction of the juice before fermentation. In this method, a portion of the sugar-rich juice is filtered with nanofiltration, ultrafiltration, or reverse osmosis membranes, which have a very small pore size and can retain the sugar. The filtered juice is then mixed with the other portion of the sugar-rich juice and fermented to obtain lower or reduced alcohol content wine [17, 18]. To produce a lower or reduced alcohol content wine ($\leq$ 10.5% v/v ethanol) by this method, optimal operating conditions and a suitable membrane configuration with a good molecular weight cutoff (MWCO) should be considered to increase the retention of volatile compounds and maintain good taste in the wine. On the contrary, this could lead to a lower content of polyphenols, anthocyanins, and color intensity and consequently affect the sensory properties of the nonalcoholic wine [17, 24].

2.1.3 Reduction of fermentable sugars using glucose oxidase

The use of glucose oxidase is another way to produce lower or reduced alcohol wines. This enzyme is found in the fungus *Aspergillus niger* and can be extracted to reduce glucose in grape juice before fermentation [19]. In this method, glucose oxidase converts β-D-glucose to D-glucono-lactone in a first reaction step reaction, releasing hydrogen peroxide, and catalyzes the conversion of D-glucono-lactone to gluconic acid in a second reaction step to produce gluconic acid. These reactions cause the oxidation of the fermentable sugars in the juice (especially glucose), which prevents the formation of ethanol from the fermentation of the sugar [25]. Treatment of a Riesling grape juice with 2 g/L glucose oxidase prior to fermentation resulted in a reduction of ethanol in the resulting wine (reduced alcohol wine) by about 4.3% v/v after 6 hours of fermentation [25]. In addition, Röcker et al. [19] achieved an ethanol reduction of 2% v/v by treating a white grape juice with glucose oxidase. This method can produce higher amounts of gluconic acid, causing the wine to become acidic and have a weak fruity odor [19, 25]. In contrast, Pickering [26] reported that the use of glucose oxidase had no significant effect on the color, aroma, flavor, and acidity of the resulting wine.

2.1.4 Reduction of fermentable sugars through early harvesting and blends

Early harvesting of fruit and blending with ripe grapes is another strategy that can be used to reduce the ethanol concentration in wine. In one study, using this strategy resulted in a 3.2% v/v reduction in ethanol content of red wines with ideal aroma profiles [27]. Similarly, a 3% v/v reduction in ethanol concentration was observed when an acidic and low alcohol blend of early harvested white and red grapes was added to a ripened grape ferment [28]. According to Piccardo et al. [29], this strategy can lower not only ethanol content but also pH and total acidity without significantly affecting other wine components. Contrary to other studies, acidity and "raw" aromas can be perceived in the resulting wine [28].

2.1.5 Reduction of fermentable sugars through viticultural practices

About half of the total fermentable sugar in grape juice is glucose [30], which is the main substrate converted to alcohol by yeasts during fermentation. Viticultural practices such as reducing photosynthetic activity, using growth regulators, reducing leaf area, and preharvest irrigation have been used to regulate grape fermentable sugars so that low-alcohol or nonalcoholic wines can be produced from grape juice [22, 23]. As indicated by some studies, the degree of sugar accumulation in berries can be influenced by reducing the leaf area [31–33], resulting in a reduction of ethanol content in the resulting wine [23, 34, 35]. For example, a lower alcohol content in the finished wine was observed after leaf area reduction of Shiraz vines [22]. A similar observation was made after post-veraison leaf removal in a Sangiovese vine, with no negative effects on phenolic compounds [36].

2.2 Reduction or limitation of alcohol production during alcoholic fermentation

Reducing or limiting alcohol production is another principle used in producing nonalcoholic or low-alcohol wines during the fermentation of wines. This principle basically includes three techniques such as interrupted fermentation [2], reduction of yeast biomass [37], use of modified yeast strains with low alcohol production ability [38, 39], and use of non-*Saccharomyces* yeasts with low alcohol production during fermentation [40, 41].

2.2.1 Reduction of alcohol production by interrupted or limited fermentation

Interrupted or limited fermentation is the intentional termination of alcoholic fermentation before it is complete by controlling the fermentation time and temperature during fermentation [2]. Generally, during fermentation, the ethanol concentration is monitored until the desired concentration is reached. Then, fermentation is stopped either by lowering the fermentation temperature or by adding sulfur dioxide. When producing nonalcoholic or low alcohol wines using this method, the fermentation time is usually short in order to achieve a very low ethanol content. However, this usually results in sweet nonalcoholic or low alcohol wines with high residual sugar content that require further post-fermentation treatments, such as heat treatment or addition of sulfur dioxide to combat microbial instability and difficult storage [42].

2.2.2 Reduction of alcohol production through yeast biomass reduction

The reduction of yeast biomass during fermentation can also be used to produce nonalcoholic or low-alcohol wines. In this method, the yeast population is reduced from time to time during fermentation to keep the fermentation rate of fermentable sugars as low as possible. This prevents the production of high amounts of ethanol during fermentation. Through centrifugation, Fan et al. [37] reduced the biomass of dry yeast (10^6 CUF/ml) during the fermentation process of an apple cider, resulting in a cider with low alcohol content and fruity aroma. Similar to limited fermentation, the final product of this technique is sweet with a high residual sugar content and requires attention for its microbial stability and storage [42]. Nevertheless, this method is useful for producing a sweeter and more pleasant nonalcoholic or low-alcohol beverage [43].

2.2.3 Reduction of alcohol production using modified yeast

The literature also reports the use of modified yeast in the production of low-alcohol and nonalcoholic wine [38, 39]. Through gene modification or adaptive evolution and selection, modified yeast strains with low ethanol production ability are developed and can be used to reduce the alcohol content in wine during fermentation [44]. A *Saccharomyces cerevisiae* strain V5 was genetically modified with an H_2O-NADH oxidase extracted from a Champagne wine yeast and used in the fermentation of a synthetic must [38]. The results showed that the modified *Saccharomyces cerevisiae* stain drastically decreased the intracellular NADH concentration and significantly altered the distribution of metabolic fluxes in the cell, resulting in the production of a lower ethanol concentration [38]. Also, genetic modification of commercial yeast strains using low-strength promoters active at different stages of fermentation regulated the expression of the *TPS1* gene, resulting in a decrease in ethanol production [39]. The problem with this technique is the release of undesirable secondary metabolites such as acetaldehyde, acetoin, and acetate [39, 45], which can negatively affect the sensory properties of the wine. In addition, the use of this technique in the production of nonalcoholic or low alcohol wines is hindered by the strong advocacy of non-GMO organisms and foods by some consumers.

2.2.4 Reduction of alcohol production using non-Saccharomyces (NS) yeasts

Non-*Saccharomyces* (NS) yeasts capable of diverting carbon or sugar metabolism to other pathways, thus preventing high ethanol production during fermentation, can be used to reduce the ethanol content of wine during fermentation[46]. Previous studies have reported the ability of NS yeast to reduce ethanol concentration within the range of 0.1–2% v/v [41, 47, 48]. For example, the sequential fermentation of *M. pulcherrima* with *S. cerevisiae* after 72 hours resulted in wines with a significant reduction in ethanol content (1.5% v/v) [49]. In addition to reducing ethanol, NS yeasts can also improve the sensory profile of wines [50–52] and control wine spoilage yeast such as *Zygosaccharomyces* species during fermentation [53].

2.3 Physical methods for removal of alcohol from wine

The complete fermentation of grape juice with high amounts of sugars produces wines. Wines are generally characterized by bitterness, hotness, good viscosity, and intense aroma and flavor. Wines can be further processed into low or nonalcoholic wines based on the final alcohol content. There are basically three methods of alcohol removal including extraction processes, membrane processes, and heat treatment [8, 9, 54].

2.3.1 Removal of alcohol by extraction processes

Extraction processes use extraction media such as gases (carbon dioxide), solvents (liquid carbon dioxide, pentane, hexane), and absorbents (zeolites) to remove ethanol from wine to produce alcohol-free or low-alcohol wine [30, 54–56]. Carbon dioxide in the form of gas or liquid can be used to extract of ethanol from wine. Carbon dioxide has a critical pressure (73 atm) and temperature (31 °C) [57], above which it behaves as a supercritical fluid (i.e., both liquid and gas) that can be used to extract organic compounds such as ethanol from wine due to its affinity for the carbon chain (as a liquid) [55] and then immediately evaporates (as a gas), leaving the extracted compound (ethanol) with a high concentration of aroma compounds [54] and no residue [7]. This method offers several advantages because carbon dioxide is inexpensive, easy to handle, does not require hazardous substances, and has a low supercritical temperature [30]. In addition, extraction solvents such as pentane and hexane are also used to remove ethanol from wines, where the ethanol dissolves in the solvent and is subsequently removed from the wine [55]. However, these extraction solvents can also remove other soluble aroma compounds along with the ethanol, which can negatively affect the aroma profile of the final product [56]. Hydrophobic adsorbents such as zeolites can also extract ethanol from wine by absorbing and filtering the ethanol from the wine. This method can be used to produce nonalcoholic wines with an ethanol content of 0.5% v/v [58]. Nevertheless, extraction methods for alcohol reduction are expensive and are rarely used in the production of low-alcohol and nonalcoholic wines.

2.3.2 Removal of alcohol by membrane processes

Membrane processes are physical separation processes that can reduce or remove ethanol from wine using a semipermeable membrane. In this method, natural osmotic pressure is created by the pressure exerted on two solutions of unequal solution concentration flowing tangentially, parallel, or circularly through a semipermeable membrane. To restore the equilibrium of natural osmotic pressure, the alcohol and water in the wine pass through the semipermeable membrane from the high-concentration solution to the low-concentration solution [30, 59]. This phenomenon reduces or removes the ethanol from the wine, resulting in a low- or non- alcoholic wine, depending on the remaining ethanol content. The most commonly used membrane separation processes at the commercial level include reverse osmosis (RO), osmotic distillation (OD), and pervaporation (PV) (**Figure 2**) [6, 8, 9, 30].

In PV, the transfer of compounds (by adsorption, diffusion, and desorption) occurs through a close-packed polymer membrane based on the partial evaporation of liquid mixtures with similar boiling points confined in an azeotropic mixture, with the liquid phase changing to the vapor phase [59, 60]. PV has been used to remove ethanol and recover aromatics from wines [61, 62]. It is highly selective, consumes little energy, operates at lower temperatures, causes less loss of aromatics, and is a clean method (i.e., it produces water and ethanol as by-products that can be recycled or reused). Nevertheless, the high cost of the PV machine and membranes, the low diffusion rate at low temperatures, and the limited market for PV membranes are some disadvantages of using PV to produce low-alcohol and nonalcoholic wines.

RO also works on the principle of membrane separation, in which a concentration gradient between two solutions through a hydrophilic, semipermeable hollow fiber membrane causes the solvent to flow from the high-concentration solution through

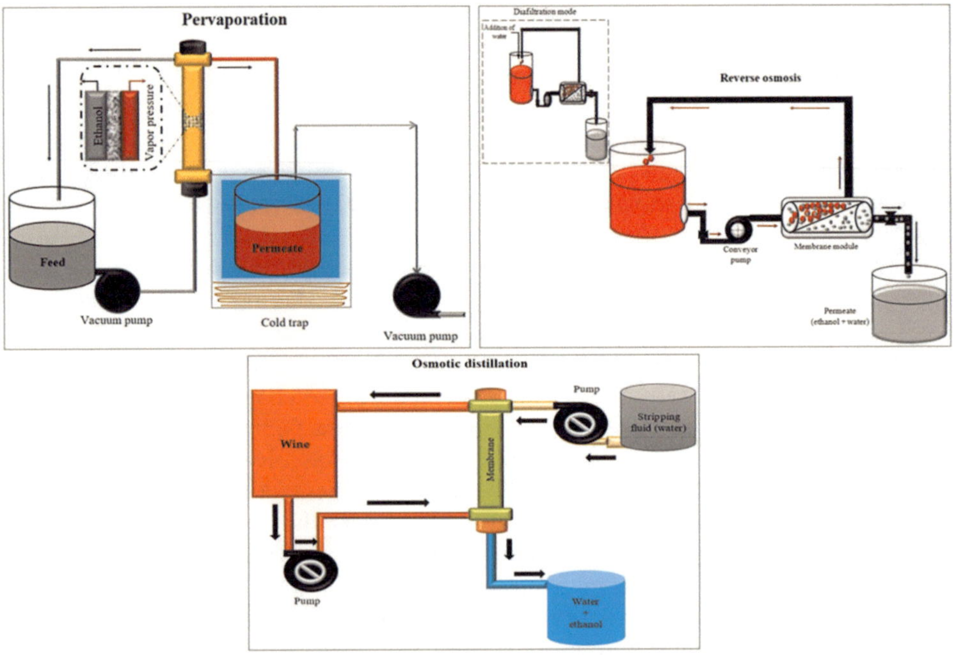

Figure 2.
Membrane separation techniques for removal of alcohol from wine.

the membrane to the low-concentration solution, retaining salts, peptides, and proteins [30]. The use of RO in the production of wines and beverages with or without alcohol content has been reported [8, 9, 63, 64]. In a diafiltration configuration, an industrial-scale plant of RO was used to produce nonalcoholic red, white, and rosé wines with a final alcohol concentration of 0.7% v/v, but most of the basic oenological parameters, volatile composition, and sensory quality of the wines were affected [65]. In contrast, some studies reported that low-alcohol or nonalcoholic wines produced with RO had no negative effects on the main aroma compounds and had similar taste and aromas to normal wines [66, 67]. RO can be operated at low temperatures and meets the requirements for a clean technology, as it can recover and reuse ethanol from the dealcoholization byproduct (water and ethanol solution). However, adding water during diafiltration to achieve effective alcohol removal is a drawback of this method, as the addition of water to wine is prohibited in some wine-producing countries or regions.

Another modern membrane separation process that has found application in the production of low-alcohol and nonalcoholic wine is osmotic distillation (OD), also known as evaporative pertraction (EP). In this membrane-based technology, two liquid phases, wine and a stripping liquid (usually water), circulate in countercurrent on opposite sides of a hydrophobic hollow fiber membrane, as shown in **Figure 2**. In this process, the vapor pressure of the volatile solutes in the wine and the stripping liquid is the driving force [68]. The mechanism for ethanol removal in the OD process is as follows: evaporation of ethanol from the wine on one side of the membrane, followed by diffusion of ethanol vapor through the membrane pores, then finally exit of ethanol vapor from the opposite side of the membrane and condensation in the stripping water solution [68]. Minimal loss of aroma compounds was observed after alcohol reduction up to 5% v/v in Garnacha, Xarelo, and Tempranillo wines by OD [69].

Similarly, alcohol reduction up to 6% v/v in fermented beverages with OD at 10 ° C–20 °C did not result in significant losses of aroma compounds [70]. Moreover, OD was used to reduce the alcohol content (−10.5% v/v) of *Montepulciano d'Abruzzo* red wine (13.23% v/v) and produce a low-alcohol wine with good aroma profile and unchanged wine color [12]. Contrarily to other studies [11, 71], adverse effects on volatiles were reported after alcohol reduction by OD.

2.3.3 Thermal (heat) processes for removing alcohol from wine

Thermal processes such as spinning cone column (SCC) and distillation under vacuum/vacuum distillation (VD) are two very common methods for reducing alcohol in wine and other alcoholic beverages based on the principle of heating and evaporation [8, 9, 30]. The SCC is a falling film separator consisting of a rotating vertical shaft and vertically stacked cones that rotate alternately and are fixed in place (**Figure 3a**). The SCC process is considered very cost-effective and efficient for retaining aroma compounds from wine, beverages, and other liquid foods [30]. In particular, it has reportedly been used to recover concentrates from grape juice, lower the ethanol content in wines, remove sulfur dioxide from grape juice, and recover aroma compounds from wines and beer [2, 9, 30, 72]. To reduce the alcohol content of the finished wine, the SCC technique uses a two-stage process. In the first stage, the wine is passed through the SCC at a reduced vacuum pressure (0.04 atm) and temperature (about 28 °C) to extract the wine aroma compounds in about 1% of the total wine volume. Subsequently, the ethanol content of the wine is reduced to produce a low-alcohol or nonalcoholic wine (depending on the final alcohol content) in the second stage at a slightly higher vacuum pressure and temperature (about 38 °C) to remove the alcohol. The aroma of the low-alcohol or alcohol-free wine is then improved by adding the recovered wine aromas (i.e., the aroma compounds extracted in the first stage) [30]. In a previous study, SCC was successfully used to recover about 97–100% of aroma compounds from white, rosé, and red wines by regulating the extraction percentage and flow rate of the base wines [72]. Moreover, 94% of ethanol was recovered from a water-ethanol mixture (14.8% v/v ethanol) using SCC at medium-high stripping rates (0.1–0.6), high feed and medium temperatures (40–50°C). However, when the alcohol content of a Chardonnay grape juice (2% v/v) was reduced halfway through fermentation with SCC, a reduction in volatile compounds

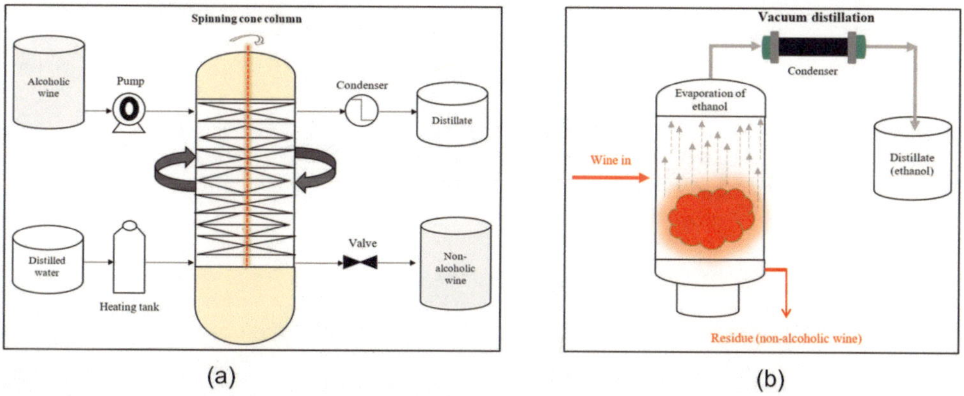

Figure 3.
Production of nonalcoholic wine using (a) spinning cone column; and (b) vacuum distillation.

of about 25% was observed. The significant change in the concentration of volatile aroma compounds after alcohol reduction could be due to the remaining ethanol content [65], the chemical-physical properties of the aroma compounds [73], and the composition of the nonvolatile matrix of the wine [74]. The high cost of SCC technology and the costs associated with its operation are two of its main drawbacks [75].

VD is another interesting technique used to reduce the alcohol content of wines and alcoholic beverages (**Figure 3b**). In this technique, the feed (usually wine) from the feed tank or flask is heated to a temperature (15–20°C) [74] suitable for the evaporation or vaporization of the ethanol of the wine from the wine medium under vacuum [72], which then condenses as a distillate in a still flask, leaving a low-alcohol or alcohol-free wine, depending on the remaining ethanol content. In some cases, some important volatile aroma compounds removed along with the ethanol could be recovered from the first distillate and added back to the nonalcoholic wine. At the same time, the ethanol could also be recovered and used for ethanol correction of wines. Previous studies have reported the use of VD to reduce the alcohol content (at 0.7–5% v/v) of wines [65, 74]. For example, the alcohol content of rosé and red wines was reduced to 5% v/v, producing reduced-alcohol wines without significantly affecting polyphenols, anthocyanins, cations, and organic acids. However, significant losses in volatile aroma compounds were observed [74]. Also, VD was used to produce nonalcoholic wines (0.7% v/v ethanol) from white, rosé, and red wines, but also significantly affected most chemical parameters and volatile composition. In particular, pH, free sulfur dioxide, total sulfur dioxide, and volatile acidity decreased significantly, while reducing sugar, color intensity, and total acidity increased significantly [65]. In addition, 92–99% of esters and terpene compounds were lost [65]. VD can significantly improve the nonvolatile components in wine compared with membrane separation methods [74]. Nevertheless, VD can significantly reduce almost all volatiles in wine, especially ethyl esters, alcohols, and terpenes [65].

Another evolving technique used in the production of low and nonalcoholic wines and beverages is an integrated membrane/distillation system known as a multistage

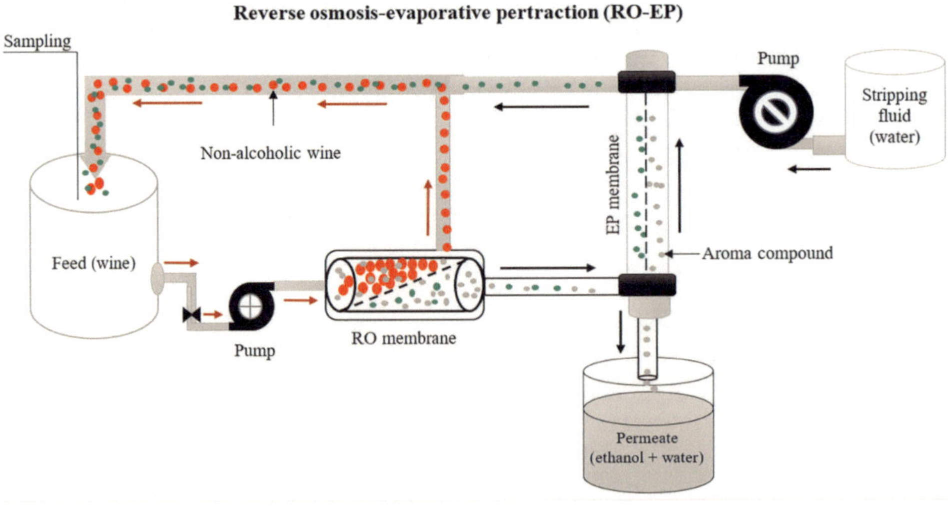

Figure 4.
Scheme of reverse osmosis-evaporative pertraction (RO-EP) for nonalcoholic wine production. RO membrane; reverse osmosis membrane and EP; and evaporative pertraction membrane.

membrane/distillation system. This technique involves the combination of two or more alcohol removal methods to remove ethanol from wines and beers while minimizing the loss of important volatile aroma compounds associated with individual membrane and thermal separation processes such as RO, PV, OD, and VD [8, 9, 12, 65, 73, 74]. Commonly used multistage membrane/distillation systems include integrated PV and distillation units [76], reverse osmosis-evaporative pertraction (RO-EP) [77], and nanofiltration-pervaporation system (NF-PV) [78], of which RO-EP is the most commonly used (**Figure 4**). These integrated systems have proven successful in producing reduced-flavor, low-alcohol, or alcohol-free wines and beers with similar or improved aroma and flavor compared with the original wine or beer product [76, 78, 79]. However, significant losses of alcohols (up to 27%), acids (up to 24%), esters (up to 22%), phenols (up to 18%), and lactones (up to 14%) have been reported at ethanol reduction up to 5.5% v/v in the case of RO-EP used for alcohol reduction of a *Montepulciano d'Abruzzo* red wine with an alcohol content of 13.2% [80].

3. Impact of production methods on the quality of low- and nonalcoholic wines

As mentioned earlier, this chapter is focused on low and nonalcoholic wines production methods used after complete fermentation (post-fermentation stage) of wine since these methods are mostly preferred to those used at pre-fermentation and concurrent fermentation stages of wine due to their ability to achieve best results, thus, produce low and nonalcoholic wines with high concentration of fermentative aroma compounds resulting from fully fermented juice. Therefore, in this section we discuss the effect of these methods on the quality of low and nonalcoholic wines, in particular, their effect on the phenolic composition, volatile compounds, and sensory characteristics.

3.1 Effect on phenolic compounds

The phenolic composition of wine (both alcoholic and non-alcoholic) is one of the key factors influencing its quality, especially red wine, which mainly includes flavonoids (anthocyanins, flavanols, flavones, flavonols, and proanthocyanidins) and non-flavonoids (hydroxybenzoic acids, hydroxycinnamic acids, and resveratrol) [81–83]. **Table 1** summarizes some reported effects of production methods on the phenolic composition of lower, reduced, low, and nonalcoholic wines. The production of nonalcoholic wines at the post-fermentation stage of winemaking using physical methods is mainly applicable to finished wines based on the principle of ethanol reduction. During the reduction of alcohol from wine, water is also removed together with the ethanol, which can have either a positive or negative effect on the phenolic composition of the final product. Wine ethanol reduction has been reported to impact wine phenolic compounds [79, 87]. The removal of ethanol from 2011 vintage Barbera red wine (14.6% v/v), Verduno Pelaverga red wine (15.2% v/v), and Langhe Rosè (13.2% v/v) by VD and membrane contactor method to a final ethanol content of 5% v/v resulted in the loss of anthocyanins and polyphenols [74]. In contrast, reduction of the alcohol level in a white wine from 12.5% v/v to 0.3% by OD had no significant effect on the total phenols and flavonoids of nonalcoholic wine compared with the base wine [11]. Similarly, flavonoids, total anthocyanins, and total phenols were not affected after the removal of ethanol (−10.5% v/v) from a red wine (13.2% v/v) [12],

Method used	Type of wine	Final alcohol content (% v/v)	Phenolic composition	Reference
OD	Montepulciano d'Abruzzo red wine	5.4	Insignificant decrease in the concentrations of total anthocyanins and total phenols	[84]
	Barbera red wine, Langhe Rosè wine, and Verduno Pelaverga red wine	5.0	Increased the concentrations of total flavonoids and total anthocyanins	[83]
	Aglianico red wine	0.4–4.9	Increased the content of total phenols	[80]
	Montepulciano d'Abruzzo red wine	2.7	Flavonoids and phenolic compounds remained unaffected	[21]
	Falanghina white wine	0.3	No significant effect on the contents of total flavonoids and total phenols	[20]
RO	Montepulciano d'Abruzzo red wine	9.0	Total anthocyanins decreased Total phenols increased	[84]
	Merlot, Cabernet Sauvignon, and Tempranillo red wines	2.0–4.0	No significant effect on the concentrations of total anthocyanins and phenolic compounds Increased color intensity by 20% due to high concentration of anthocyanins	[85]
SCC	White, rose, and red wines	< 0.3	Increased the contents of flavonols, anthocyanins, total phenols, and phenolic compounds contents by 24%	[86]
VD	Langhe Rosè wine, Verduno Pelaverga red wine, and Barbera red wine	5.0	Increased the contents of total anthocyanins and total flavonoids	[83]
RO–EP	Montepulciano d'Abruzzo red wine (cv.)	5.5	Increased the content of total phenols Decreased the content of total anthocyanins	[84]

Table 1.
Some reported effects of production methods on the phenolic composition of lower, reduced, low, and nonalcoholic wines.

whereas a reduction of up to 5% v/v ethanol in a red wine by RO-EP caused an increase in the color intensity and phenolic compounds concentration [88]. Furthermore, SCC was reported to modify the phenolic composition of red wines reduced to less than 0.3 % v/v such that the concentrations of phenolic compounds including total phenols, anthocyanins, stilbenes, flavonols, flavan-3-ols, and non-flavonoids increased significantly [87].

3.2 Effect on volatile components

The aroma and flavor of wines are mainly associated with volatile aroma compounds belonging to different chemical groups such as esters, organic acids, alcohols, terpenes, monoterpenes, C_{13} norisoprenoids, aldehydes, ketones, lactones, and sulfur compounds [89]. These compounds are either of varietal (imparted from the grape skins), fermentative (produced during wine fermentation) or post-fermentative (produced from aging or additives after fermentation) origin. Factors such as grape variety, viticultural practices, and winemaking methods define the volatile composition of wines as well as its aroma and flavor [90]. As regards the production of lower, reduced, low, and nonalcoholic wines using post-fermentation methods such as membrane separation and heat treatment processes, the removal of alcohol can affect the volatile compounds of the final product. For example, the total removal of ethanol from a Tokaji Hárslevelű wine with an alcohol content of 13.1% v/v by PV resulted in a 70% loss of the total aroma compounds [61]. In addition, the production of a nonalcohol wine (0.5% v/v) from a Cabernet Sauvignon red wine (12.5% v/v) using PV led to losses of 99, 28, and 40% of esters, organic acids, and alcohols, respectively [62]. Furthermore, losses of about 9, 4, and 18% were observed in the total concentration of volatile compounds in white, rose, and red wines, respectively, after treated with SCC [72]. More recently, Sam et al. [65] compared RO and VD in the obtainment of nonalcoholic wines (with final ethanol content of 0.7 v/v) from white wine (13.4% v/v), rose wine (12.2% v/v), and red wine (13.9% v/v). They observed significant losses of volatile compounds in the nonalcoholic wines, in particular, VD resulted in losses of the total concentration of esters in white, rosé, and red wines by 96, 98, and 96%, respectively, whereas respective losses of 92, 81, and 87% were observed in RO-treated wines. Alcohol removal is not solely responsible for the losses of volatile aroma compounds during the production of lower, reduced, low, and nonalcoholic wines, other factors such as the type of method used, the operating conditions applied, the type of membrane used (in the case of membrane processes), the chemical-physical properties of the volatile compounds, and the nonvolatile matrix of the wine can also play a vital role [8, 9]. Some reported effects of production methods on the volatile compounds of lower, reduced, low, and nonalcoholic wines are summarized in **Table 2**.

3.3 Effect on sensory characteristics

Volatile compounds, especially terpenes and esters, contribute significantly to the aroma and flavor of wines [94, 95], and their complete loss or decrease due to the removal of ethanol from wine can significantly affect the sensory characteristics of the final wine product. Ethanol can enhance the perception of viscosity, bitterness, and hotness in wine, while masking other sensory characteristics such as astringency and acidity [85, 86, 96]. Some important findings on the effect of ethanol reduction using nonalcoholic wines production methods are presented in **Table 3**. Studies have shown that the production of lower, reduced, low, and nonalcoholic wines by post-fermentation techniques can significantly affect sensory attributes such as hotness, bitterness, aroma intensity, color, astringency, acidity, sweetness, wine body, red fruits, dried fruits, etc. [12, 62, 65, 71, 91, 99, 100]. A nonalcoholic white wine (0.3% v/v ethanol) produced by OD was characterized by low sweetness, aroma, viscosity, and high acidity in comparison to the original the wine with an alcohol content of 12.5%, giving it an unbalanced taste and unpleasant aftertaste [11]. Similar observations were made in nonalcohol white, rose, and red wines produced by RO and VD [65]. Moreover, the reduction of ethanol in Aglianico red wines at 5% v/v by a membrane contactor technique decreased aroma notes such as red fruits and cherry in the final reduced wine products [88]. Furthermore, when SCC was used to reduce the alcohol content of oaked Chardonnay wine, the perceptions of hotness and overall aroma intensity reduced substantially compared with the original wine [92]. It is worth mentioning that low and nonalcoholic wines (< 0.5–5.5% v/v ethanol) usually

Method used	Type of wine	Final alcohol content (% v/v)	Losses of volatile compounds (%)	Reference
OD	Aglianico red wine	8.8	Esters = 60.9 Alcohols = 31.8 Acids = 17.1 Terpene compounds = 32.3	[91]
	Montepulciano d'Abruzzo red wine	5.4	Esters = 19.0 Alcohols = 3.0 Acids = 25.0 lactones = 25.0 Phenols = 10.0	[84]
	Langhe Rosè wine, Barbera red wine, and Verduno Pelaverga red wine	5.0	Esters = 23.8–47.8 Alcohols = 59.9–63.9 Acids = 17.4–30.9	[83]
	Montepulciano d'Abruzzo red wine	2.7	Esters = 85.0 Alcohols = 84.0 Acids = 23.0 lactones = 37.0 Phenols = 37.0	[21]
	Falanghina white wine	0.3	Esters = 99.0 Alcohols = 98.9 Acids = 98.7 Ketones = 99.9 Lactones = 98.2	[20]
	Aglianico red wine	0.2	Esters = 89.9 Alcohols = 99.9 Acids = 78.9 Ketones & lactones = 97.9 Aldehydes = 100 Sulfur compounds = 78.7 Phenols = 100	[82]
RO	Montepulciano d'Abruzzo red wine	9.0	Esters = 8.0 Alcohols = 30.0 Acids = 22.0 Phenols = 13.0 Lactones = 14.0	[84]
	Chardonnay white wine, Pinot Noir rose wine, and Merlot red wine	0.7	Esters = 81–92 Alcohols = 58–75 Acids = 73–89 Terpenes = 48–70 Other compounds = 75–87	[74]
PV	Cabernet Sauvignon red wine	0.5	Esters = 99.9 Alcohols = 39.5 Acids = 28.2	[71]
SCC	White wine	0.3	Esters = 53.0 Aliphatic alcohols = 98.0 Aromatic alcohols = 3.0 Acids = 20.0 Ketones = 71.0	[92]
VD	Chardonnay white wine, Pinot Noir rose wine, and Merlot red wine	0.7	Esters = 96–98 Alcohols = 85–95 Acids = 85–91	[74]

Method used	Type of wine	Final alcohol content (% v/v)	Losses of volatile compounds (%)	Reference
			Terpene compounds = 92–96 Other compounds = 91–99	
	Langhe Rosè wine, Barbera red wine, and Verduno Pelaverga red wine	5.0	Esters = 19.5–22.9 Alcohols = 50.4–53.6 Acids = 2.3–13.7	[83]
RO–EP	Shiraz red wine	10.4	Esters = 49.5 Alcohols = 38.9 Terpene compounds = 35.3 Lactones = 21.4	[93]
	Montepulciano d'Abruzzo red wine	5.5	Esters = 22.0 Alcohols = 27.0 Acids = 24.0 Phenols = 18.0 Lactones = 14.0	[84]

Table 2.
Some reported effects of production methods on the volatile compounds of lower, reduced, low, and nonalcoholic wines.

have poor sensory quality and consumer preferences due to their imbalanced body and flavor, reduced hotness, and high acidity and astringency when compared with original wines [65, 91, 99] unless supplemented with additives. Meanwhile, lower and reduced wines (6.5–10.5% v/v ethanol) typically have acceptable preferences [12, 88, 93] due to less negative impact on the sensory characteristics arising from less ethanol removal and aroma compounds. For example, the reduction of alcohol in a white Chardonnay wine (14.2% v/v) by 4.5% v/v negatively affected consumer liking of the final product, while a reduction of 1.5% and 3.3% v/v had no significant effect. Also, when a red wine with alcohol content of 13.2% v/v was dealcoholized (i.e., its ethanol reduced) by 8% v/v, no substantial changes in the color intensity and overall acceptability were observed between the two wines [12]. In addition, an ethanol reduction by 3 and 5% v/v in two red wines (cv. Aglianico) with different initial alcohol contents (15.4 and 13.3 % v/v) using a membrane contactor technique increased the bitterness, acidity, and astringency of the final lower alcohol wines, while a 2% v/v reduction resulted in no significant differences between the base wines and the final wine products [88]. Similarly, Meillon et al. [93] reported a decrease in consumer preference for a Syrah red wine (13.4% v/v) dealcoholized by 5.5% v/v and a nonsignificant effect on the preference at dealcoholization by 2% and 4% v/v using RO. The inability of most consumers to notice alcohol reductions ≤2% v/v may have accounted for these results [8].

4. Aroma improvement of lower, reduced, low, and nonalcoholic wines

The aroma profiles of lower, reduced, low, and non-alcoholic wines have a great impact on consumers' acceptability and mostly depend on volatile compounds. As the removal of alcohol from finished wines usually results in substantial loss of volatile compounds leading to changes in organoleptic properties, innovative ways for correcting these adverse effects are needed. Ways of improving the aroma and

Method used	Type of wine	Final alcohol content (% v/v)	Sensory characteristics	Reference
OD	Aglianico red wine	8.8–11.6	Astringency, bitterness, and acidity increased, while red fruits, sweet, and cherry aromas decreased.	[91]
	Falanghina white wine	0.3–9.8	Unbalanced taste and liking, with an unpleasant aftertaste due to reduced sweetness, body, and odor	[20]
	Montepulciano d'Abruzzo red wine	2.7–8.3	Lower acceptability due to high acidity, low sweetness, and low red fruits and spices notes	[21]
RO	Syrah red wine	9.6	Hotness, sweetness, and wine length in the mouth decreased, while red fruits, woody and blackcurrant aromas increased	[97]
	Syrah red wine	7.9	Aromas, persistence, and complexity decreased	[98]
	Chardonnay white wine, Pinot Noir rose wine, and Merlot red wine	0.7	Acidity, astringency, and color intensity increased, while viscosity, sweetness, fruity and floral notes, red fruits, bitterness, hotness, aroma intensity, and overall acceptability decreased	[74]
PV	Cabernet Sauvignon red wine	0.5	Good smell and taste due to high retention of fruity aromas	[71]
VD	Chardonnay white wine, Pinot Noir rose wine, and Merlot red wine	0.7	Acidity, astringency, and color intensity increased, while viscosity, sweetness, fruity and floral notes, red fruits, bitterness, hotness, aroma intensity, and overall acceptability decreased	[74]
RO–EP	Shiraz red wine	10.4	Dark fruit, raisin/prune, and black pepper notes increased. Astringency and overall aroma intensity also Increase	[93]

Table 3.
Some reported effects of production methods on the sensory properties of lower, reduced, low, and nonalcoholic wines.

sensory properties of lower, reduced, low, and nonalcoholic wines are rarely reported in the literature although the use of fruit-based, herb-based, and other rarely used aroma additives in enhancing the aroma profile of wines and alcoholic beverages has been reported [101]. For example, the concentration of monoterpenes and monoterpene glycosides significantly increased after the addition of phenolic-free glycosides, resulting in an increase in floral and fruity aroma, flavor, and aftertaste attributes, without altering the bitterness or astringency [102]. Similarly, dehydrated waste grape skins were used to improve the aroma composition of red wines [103]. Furthermore, addition of hydroalcoholic plants macerates to Vermouth and basic wines improved their sensory characteristics such as aroma, taste, and smell [104]. Moreover, when 2 g/L of Ganoderma lucidum extract was added to a Shiraz wine product, it imparted the wine with fruity and floral notes [105]. Also, the addition of oak chips to Verdejo wines imparted the wines with higher concentrations ethyl acetate, hexyl acetate, isoamyl acetates, higher alcohols, and ethyl esters of straight-chain fatty acids [106], which are known to contribute fruity and floral aromas to the wines. Other wines aromatized with botanical extracts include Benedictine, Chartreuse, liqueurs, and bitters [97]. Possibly, the reciprocation of these studies in lower, reduced, low, and nonalcoholic wines would significantly improve their aroma profile. However, the ongoing debate at OIV whether to permit the use of flavorings or exogenous aroma additives from grapes or non-grapes origin in the aroma improvement of these categories of products is a major hindrance to the scientific exploration in this field. Nevertheless, some studies have reported on the aroma improvement of dealcoholized wines (lower, reduced, low, and nonalcoholic wines) and beers [11, 76, 107]. In an attempt to improve the aroma profile of a white wine (11.5% v/v) dealcoholized to a final ethanol content of 0.8% v/v by vacuum evaporation, glycosidic aroma precursors isolated from Muscat grapes were added to the dealcoholized wine. This increased concentrations of β-phenylethyl alcohol, linalool, and geraniol, imparting the final product with high fruity and floral odors [107]. Similarly, Liguori et al. [11] developed an alcohol-free wine beverage with improved aftertaste and flavor from an OD dealcoholized white wine (0.3% v/v ethanol) by adding grape must, sodium carbonate solution, and some floral wine flavors. Furthermore, the aroma profile of a nonalcoholic beer with alcohol content of less than 0.5% was improved by first extracting aroma compounds from non-carbonated alcoholic beer (5.67% v/v ethanol) by pervaporation. Subsequently, the alcohol was removed from the alcoholic beer by spinning cone column distillation. The dealcoholized beer was then reconstituted with about 5–10% v/v of the original beer and 0.3% v/v of the extracted aroma compounds and finally carbonated, resulting in a nonalcoholic beer with improved aroma profile similar to the original beer [76]. Recently, a study was conducted at Gansu Key Laboratory of Viticulture and Enology, College of Food Science and Engineering, Gansu Agricultural University, China to investigate the effect of rose (*R. chinensis* var. spontanea red) peach (*Prunus persica*), and lily (*Lilium bulbiferum*) flower extracts on the aroma profile of dealcoholized red and rose wines (0.7% v/v ethanol). The dealcoholized wines were reconstituted by the addition of the flower extracts. Sensory analysis was performed, which revealed that the aroma profile of the reconstituted

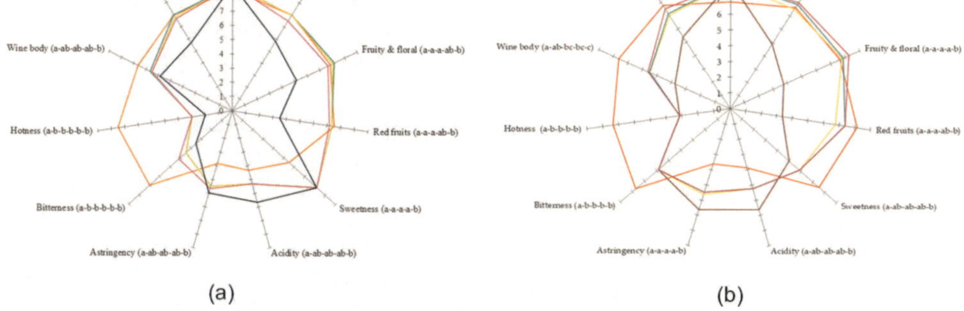

Figure 5.
Spider plot of sensory analysis (means) performed on (a) rose wine; and (b) red wine. Different letters (a–c) represent significant differences at a significant level of 0.05. OW; original merlot wine (control), DW; dealcoholized merlot wine (control), R-RDW; rose reconstituted dealcoholized wine, P-RDW; peach reconstituted dealcoholized wine, L-RDW; and lily reconstituted dealcoholized wine.

dealcoholized wines improved significantly after the addition of the extracts compared with the dealcoholized wines. In particular, the aroma attributes such as red fruits, fruity and floral, and aroma intensity increased (**Figure 5**), which was attributed to some aroma compounds including isoamyl acetate, ethyl hexanoate, ethyl octanoate, isoamyl octanoate, phenethyl acetate, linalool, β-damascenone, and geraniol imparted by the added flower extracts. These aroma compounds are known to contribute fruity and floral aromas to wine [98, 108–110]. In addition, the reconstituted dealcoholized wines were perceived sweeter and less acidic and astringent with improved wine body and overall acceptability among the panelists (**Figure 5**).

5. Perspectives

With health risks awareness, consumer preferences are shifting toward new product offerings and alternatives, with increasing percentage of the adult population seeking lower alcohol wines more frequently. This has boosted nonalcoholic wine production and sales, with many industries and researchers already abreast with different nonalcoholic wine production techniques at the various stages of winemaking. In this chapter, we focus on the methods used for the production of NW from high-strength alcoholic wines after complete fermentation (wine post-fermentation stage). Specifically, their impact on the aroma profile and sensory characteristics of NW as well as the state-of-the-art methods of improving the aroma profile of such product. Among the methods of NW production, physical dealcoholization methods are usually used as they can achieve the best results when used on a finished wine. Also, when used in the reduction of ethanol at several percent (2–4% v/v), they can preserve the phenolic compounds, volatile compounds, and sensory quality of the wine. Furthermore, the end product usually has a taste almost similar to original wine. In contrast, the other methods discussed in this chapter can produce unbalanced wines (with high acidity, unfermented juice, and low fermentative aroma compounds) with legality issues in the case of the juice fermentable sugars dilution with water. Nevertheless, some important aroma compounds can be lost using physical dealcoholization methods in the production of NW. Therefore, subsequent aroma enhancement may be needed to compensate for the loss of important volatile compounds associated with the aroma profile of the NW during dealcoholization. Currently, there are few studies that scientifically evaluate or optimize the parameters of the production process of aroma-enhanced dealcoholized wines, which could be one of the future research areas. To date, there is limited research on new types of aroma-enhanced dealcoholized wines, though there is evidence that the use of fruit-based, herb-based, and other rarely used aromatic materials in winemaking improves the aroma profiles of wines and dealcoholized wines. Moreover, the unapproved use of fruit-based, herb-based, and other aromatic materials as an oenological practice by the European Union (EU) and the International Organization of Vine and Wine (OIV) is a major setback to their use as wine additives. Nevertheless, for the category of special and aromatized wines, they could be added. The development of novel products from dealcoholized wines reconstituted with fruit-based, herb-based, or new aroma additives represents a potential new market for the wine industry. Therefore, future development of such products will benefit not only the wine industry by producing diversified and high-quality commercial NW and wine products, but also consumers by providing options for novel aroma-enhanced dealcoholized wines with unique and pleasant aroma profiles.

Acknowledgements

The authors would like to thank the Department of Commerce of Gansu Province fund, through the wine style exploration, solidification, and typical new product development in Gansu Hexi Corridor appellation, project: GSPTJZX-2020-4, and the Ministry of Education of Gansu Province fund, through the wine production practice teaching reform and practice based on industry chain system, project: CXCYJG2021003. We would also like to thank the late professor Shun-Yu Han from the College of Food Science and Engineering, Gansu Agricultural University, and a former Dean of Gansu Key Laboratory of Viticulture and Enology for his help in fund acquisition and his contribution in the conception of this chapter, and express our gratitude to professor Yong-Cai Li, the dean of College of Food Science and Engineering, Gansu Agricultural University for his support in writing this book chapter.

Conflict of interest

The authors declare no conflict of interest.

Author details

Teng-Zhen Ma*, Faisal Eudes Sam and Bo Zhang
Gansu Key Laboratory of Viticulture and Enology, College of Food Science and Engineering, Gansu Agricultural University, Lanzhou, China

*Address all correspondence to: matz@gsau.edu.cn

References

[1] Wines. GB 15037-2006; the State Standard of the People's Republic of China. Beijing, China: Standards Press of China; 2006

[2] Pickering GJ. Low and Reduced-alcohol Wine : A Review. Journal of Wine Research. 2000;**11**:37-41. DOI: 10.1080/0957126002000157

[3] Saliba AJ, Ovington LA, Moran CC. Consumer demand for low-alcohol wine in an Australian sample. International Journal of Wine Business Research. 2013; **5**:1-8. DOI: 10.2147/IJWR.S41448

[4] Wine Australia. Low Alcohol Wine; Wine Australia: Adelaide, Australia, 2021

[5] Report FACT4532MR. Non-Alcoholic Wine Market Forecast, Trend Analysis & Competition Tracking. London, UK: Global Market Insights 2019 to 2027; Market Forecast R; 2020

[6] Liguori L, Russo P, Albanese D, Di Matteo M. Production of Low-Alcohol Beverages: Current Status and Perspectives. Amsterdam, The Netherlands: Elsevier Inc.; 2018

[7] Mangindaan D, Khoiruddin K, Wenten IG. Beverage dealcoholization processes: Past, present, and future. Trends in Food Science and Technology. 2018;**71**:36-45

[8] Longo R, Blackman JW, Torley PJ, Rogiers SY, Schmidtke LM. Changes in volatile composition and sensory attributes of wines during alcohol content reduction. Journal of the Science of Food and Agriculture. 2017;**97**:8-16. DOI: 10.1002/jsfa.7757

[9] Sam FE, Ma T, Salifu R, Wang J, Jiang Y-M, Zhang B, et al. Techniques for dealcoholization of wines: Their impact on wine phenolic composition, volatile composition, and sensory characteristics. Food. 2021;**10**:2498. DOI: 10.3390/foods10102498

[10] Banvolgyi S, Savaş Bahçeci K, Vatai G, Bekassy S, Bekassy-Molnar E. Partial dealcoholization of red wine by nanofiltration and its effect on anthocyanin and resveratrol levels. Food Science and Technology International. 2016;**22**:677-687. DOI: 10.1177/1082013216642331

[11] Liguori L, Albanese D, Crescitelli A, Di Matteo M, Russo P. Impact of dealcoholization on quality properties in white wine at various alcohol content levels. Journal of Food Science and Technology. 2019;**56**:3707-3720. DOI: 10.1007/s13197-019-03839-x

[12] Corona O, Liguori L, Albanese D, Di Matteo M, Cinquanta L, Russo P. Quality and volatile compounds in red wine at different degrees of dealcoholization by membrane process. European Food Research and Technology. 2019;**245**: 2601-2611. DOI: 10.1007/s00217-019-03376-z

[13] Fedrizzi B, Nicolis E, Camin F, Bocca E, Carbognin C, Scholz M, et al. Stable Isotope Ratios and Aroma Profile Changes Induced Due to Innovative Wine Dealcoholisation Approaches. Food and Bioprocess Technology. 2014;7:62-70. DOI: 10.1007/s11947-013-1068-x

[14] Teng B, Petrie PR, Nandorfy DE, Smith P, Bindon K. Pre-Fermentation water addition to high-sugar shiraz must: Effects on wine composition and sensory properties. Foods. 2020;**9**:1193. DOI: 10.3390/foods9091193

[15] Teng B, Petrie PR, Smith PA, Bindon KA. Comparison of water addition and early-harvest strategies to decrease alcohol concentration in Vitis vinifera cv. Shiraz wine: impact on wine phenolics, tannin composition and colour properties. Australian Journal of Grape and Wine Research. 2020;**26**:158-171. DOI: 10.1111/ajgw.12430

[16] Schelezki OJ, Deloire A, Jeffery DW. Substitution or dilution? Assessing pre-fermentative water implementation to produce lower alcohol Shiraz wines. Molecules. 2020;**25**:2245. DOI: 10.3390/molecules25092245

[17] Mira H, Guiomar A, Geraldes V, De Pinho MN. Membrane processing of grape must for control of the alcohol content in fermented beverages. Journal of Membrane Science and Research. 2017;**3**:308-312. DOI: 10.22079/JMSR.2017.60634.1130

[18] Salgado CM, Fernández-Fernández E, Palacio L, Hernández A, Prádanos P. Alcohol reduction in red and white wines by nanofiltration of musts before fermentation. Food and Bioproducts Processing. 2015;**96**:285-295. DOI: 10.1016/j.fbp.2015.09.005

[19] Röcker J, Schmitt M, Pasch L, Ebert K, Grossmann M. The use of glucose oxidase and catalase for the enzymatic reduction of the potential ethanol content in wine. Food Chemistry. 2016;**210**:660-670. DOI: 10.1016/j.foodchem.2016.04.093

[20] Ruiz E, Busto MD, Ramos-Gómez S, Palacios D, Pilar-Izquierdo MC, Ortega N. Encapsulation of glucose oxidase in alginate hollow beads to reduce the fermentable sugars in simulated musts. Food Bioscience. 2018;**24**:67-72. DOI: 10.1016/j.fbio.2018.06.004

[21] Schelezki OJ, Smith PA, Hranilovic A, Bindon KA, Jeffery DW. Comparison of consecutive harvests versus blending treatments to produce lower alcohol wines from Cabernet Sauvignon grapes: Impact on polysaccharide and tannin content and composition. Food Chemistry. 2018;**244**:50-59. DOI: 10.1016/j.foodchem.2017.10.024

[22] Zhang P, Wu X, Needs S, Liu D, Fuentes S, Howell K. The influence of apical and basal defoliation on the canopy structure and biochemical composition of Vitis vinifera cv. shiraz grapes and wine. Frontiers in Chemistry. 2017;**5**:1-9. DOI: 10.3389/fchem.2017.00048

[23] Sun Q, Sacks GL, Lerch SD, Vanden Heuvel JE. Impact of shoot and cluster thinning on yield, fruit composition, and wine quality of Corot noir. American Journal of Enology and Viticulture. 2012;**63**:49-56. DOI: 10.5344/ajev.2011.11029

[24] García-Martín N, Perez-Magariño S, Ortega-Heras M, González-Huerta C, Mihnea M, González-Sanjosé ML, et al. Sugar reduction in musts with nanofiltration membranes to obtain low alcohol-content wines. Separation and Purification Technology. 2010;**76**:158-170. DOI: 10.1016/j.seppur.2010.10.002

[25] Pickering G. The Production of Reduced-Alcohol Wine Using Glucose Oxidase. Ph.D. Thesis, Lincoln, UK: Lincoln University; 1997

[26] Pickering GJ, Heatherbell DA, Barnes MF. The production of reduced-alcohol wine using glucose oxidase-treated juice. Part III. Sensory. American Journal of Enology and Viticulture. 1999;**50**:307-316

[27] Asproudi A, Ferrandino A, Bonello F, Vaudano E, Pollon M, Petrozziello M. Key norisoprenoid

compounds in wines from early-harvested grapes in view of climate change. Food Chemistry. 2018;**268**: 143-152. DOI: 10.1016/j.foodchem.2018.06.069

[28] Balda P, Martinez de Toda F. Decreasing the alcohol level in quality red wines by the "double harvest" technique. In: 17 th International Symposium Giesco 2011. Italia: Asti; 2011

[29] Piccardo D, Favre G, Pascual O, Canals JM, Zamora F, González-Neves G. Influence of the use of unripe grapes to reduce ethanol content and pH on the color, polyphenol and polysaccharide composition of conventional and hot macerated Pinot Noir and Tannat wines. European Food Research and Technology. 2019;**245**:1321-1335. DOI: 10.1007/s00217-019-03258-4

[30] Schmidtke LM, Blackman JW, Agboola SO. Production technologies for reduced alcoholic wines. Journal of Food Science. 2012;**77**:25-41. DOI: 10.1111/j.1750-3841.2011.02448.x

[31] Parker AK, Hofmann RW, van Leeuwen C, Mclachlan ARG, Trought MCT. Manipulating the leaf area to fruit mass ratio alters the synchrony of total soluble solids accumulation and titratable acidity of grape berries. Australian Journal of Grape and Wine Research. 2015;**21**: 266-276. DOI: 10.1111/ajgw.12132

[32] De Toda MF, Balda P. Delaying berry ripening through manipulating leaf area to fruit ratio. Vitis—Journal of Grapevine Research. 2013;**52**: 171-176

[33] Palliotti A, Cini R, Leoni F, Silvestroni O, Poni S. Effects of late mechanized leaf removal above the clusters zone to delay grape ripening in "Sangiovese" vines. Acta Horticulturae. 2013;**978**:301-308. DOI: 10.17660/actahortic.2013.978.35

[34] De Toda FM, Sancha JC, Balda P. Reducing the sugar and pH of the grape (vitis vinifera l. Cvs. "Grenache" and 'tempranillo') through a single shoot trimming. South African Journal of Enology and Viticulture. 2013;**34**: 246-251. DOI: 10.21548/34-2-1101

[35] Palma LD, Tarricone L, Muci G, Limosani P, Savino M, Novello V. Leaf removal, vine physiology and wine quality in cv. Negroamaro (Vitis vinifera L.). Le Progres Agricole Et Viticole. 2011; **17**:231-234

[36] Poni S, Gatti M, Bernizzoni F, Civardi S, Bobeica N, Magnanini E, et al. Late leaf removal aimed at delaying ripening in cv. Sangiovese: Physiological assessment and vine performance. Australian Journal of Grape and Wine Research. 2013;**19**:378-387. DOI: 10.1111/ajgw.12040

[37] Fan G, Shengyun T, Rong W, Qingbin L, Jinsheng Z, Xiaodong Y, et al. BioengineeringDepartment Fermentation process of low-alcohol cider by biomass reduction. China Brew. 2012;**31**:186-190

[38] Heux S, Sablayrolles JM, Cachon R, Dequin S. Engineering a Saccharomyces cerevisiae wine yeast that exhibits reduced ethanol production during fermentation under controlled microoxygenation conditions. Applied and Environmental Microbiology. 2006; **72**:5822-5828. DOI: 10.1128/AEM.00750-06

[39] Rossouw D, Heyns EH, Setati ME, Bosch S, Bauer FF. Adjustment of trehalose metabolism in wine saccharomyces cerevisiae strains to modify ethanol yields. Applied and

Environmental Microbiology. 2013;**79**: 5197-5207. DOI: 10.1128/AEM.00964-13

[40] Lemos Junior WJF, Nadai C, Crepalde LT, de Oliveira VS, Dupas de Matos A, Giacomini A, et al. Potential use of Starmerella bacillaris as fermentation starter for the production of low-alcohol beverages obtained from unripe grapes. International Journal of Food Microbiology. 2019;**303**:1-8. DOI: 10.1016/j.ijfoodmicro.2019.05.006

[41] Canonico L, Solomon M, Comitini F, Ciani M, Varela C. Volatile profile of reduced alcohol wines fermented with selected non-Saccharomyces yeasts under different aeration conditions. Food Microbiology. 2019;**84**:103247. DOI: 10.1016/j.fm.2019.103247

[42] Malfeito-Ferreira M. Yeasts and wine off-flavours: A technological perspective. Annales de Microbiologie. 2011;**61**:95-102. DOI: 10.1007/s13213-010-0098-0

[43] Nogueira A, Mongruel C, Simões DRS, Waszczynskyj N, Wosiacki G. Effect of biomass reduction on the fermentation of cider. Brazilian Archives of Biology and Technology. 2007;**50**:1083-1092. DOI: 10.1590/s1516-89132007000700019

[44] Tilloy V, Cadière A, Ehsani M, Dequin S. Reducing alcohol levels in wines through rational and evolutionary engineering of Saccharomyces cerevisiae. International Journal of Food Microbiology. 2015;**213**:49-58. DOI: 10.1016/j.ijfoodmicro.2015.06.027

[45] Varela C, Kutyna DR, Solomon MR, Black CA, Borneman A, Henschke PA, et al. Evaluation of gene modification strategies for the development of low-alcohol-wine Yeasts. Applied and Environmental Microbiology. 2012; **78**:6068-6077. DOI: 10.1128/AEM.01279-12

[46] Kutyna DR, Varela C, Henschke PA, Chambers PJ, Stanley GA. Microbiological approaches to lowering ethanol concentration in wine. Trends in Food Science and Technology. 2010;**21**: 293-302. DOI: 10.1016/j.tifs.2010.03.004

[47] Al Daccache M, Koubaa M, Salameh D, Vorobiev E, Maroun RG, Louka N. Control of the sugar/ethanol conversion rate during moderate pulsed electric field-assisted fermentation of a Hanseniaspora sp. strain to produce low-alcohol cider. Innovative Food Science and Emerging Technologies. 2020;**59**: 102258. DOI: 10.1016/j.ifset.2019.102258

[48] Ciani M, Morales P, Comitini F, Tronchoni J, Canonico L, Curiel JA, et al. Non-conventional yeast species for lowering ethanol content of wines. Frontiers in Microbiology. 2016;**7**:1-13. DOI: 10.3389/fmicb.2016.00642

[49] Canonico L, Comitini F, Oro L, Ciani M. Sequential fermentation with selected immobilized non Saccharomyces yeast for reduction of ethanol content in wine. Frontiers in Microbiology. 2016;**7**:278. DOI: 10.3389/fmicb.2016.00278

[50] Du Plessis H, Du Toit M, Nieuwoudt H, Van Der Rijst M, Kidd M, Jolly N. Effect of saccharomyces, non-saccharomyces yeasts and malolactic fermentation strategies on fermentation kinetics and flavor of Shiraz wines. Fermentation. 2017;**3**:1-24. DOI: 10.3390/fermentation3040064

[51] Varela C, Barker A, Tran T, Borneman A, Curtin C. Sensory profile and volatile aroma composition of reduced alcohol Merlot wines fermented with Metschnikowia pulcherrima and Saccharomyces

uvarum. International Journal of Food Microbiology. 2017;**252**:1-9. DOI: 10.1016/j.ijfoodmicro.2017.04.002

[52] Maturano YP, Mestre MV, Kuchen B, Toro ME, Mercado LA, Vazquez F, et al. Optimization of fermentation-relevant factors: A strategy to reduce ethanol in red wine by sequential culture of native yeasts. International Journal of Food Microbiology. 2019;**289**:40-48. DOI: 10.1016/j.ijfoodmicro.2018.08.016

[53] Alonso A, Belda I, Santos A, Navascués E, Marquina D. Advances in the control of the spoilage caused by Zygosaccharomyces species on sweet wines and concentrated grape musts. Food Control. 2015;**51**:129-134. DOI: 10.1016/j.foodcont.2014.11.019

[54] Ruiz-rodríguez A, Fornari T, Jaime L, Vázquez E, Amador B, Antonio J, et al. The Journal of Supercritical Fluids Supercritical CO 2 extraction applied toward the production of a functional beverage from wine. Journal of Supercritical Fluids. 2012;**61**:92-100. DOI: 10.1016/j.supflu.2011.09.002

[55] Trifunovic O, Trägårdh G. The influence of permeant properties on the sorption step in hydrophobic pervaporation. Journal of Membrane Science. 2003;**216**:207-216. DOI: 10.1016/S0376-7388(03)00072-3

[56] Raventós M, Duarte S, Alarcón R. Application and Possibilities of Supercritical CO2 Extraction in Food Processing Industry: An Overview. Food Science and Technology International. 2002;**8**:269-284. DOI: 10.1106/108201302029451

[57] Atkins P, de Paula J. Atkins. Physical Chemistry Chemical Physics. 2006;**8**:798

[58] Akyereko YG, Wireko-manu FD, Alemawor F, Adzanyo M. Nonalcoholic Wine. Vol. 20212021. p. 10

[59] Wenten IG, Dharmawijaya PT, Aryanti PTP, Mukti RR. Khoiruddin LTA zeolite membranes: Current progress and challenges in pervaporation. RSC Advances. 2017;**7**:29520-29539. DOI: 10.1039/c7ra03341a

[60] Crespo JG, Brazinha C. 1—Fundamentals of Pervaporation. In Pervaporation, Vapour Permeation and Membrane Distillation. Cambridge, UK: Woodhead Publishing; 2015. pp. 3-17

[61] Takács L, Vatai G, Korány K. Production of alcohol free wine by pervaporation. Journal of Food Engineering. 2007;**78**:118-125. DOI: 10.1016/j.jfoodeng.2005.09.005

[62] Sun X, Dang G, Ding X, Shen C, Liu G, Zuo C, et al. Production of alcohol-free wine and grape spirit by pervaporation membrane technology. Food and Bioproducts Processing. 2020; **123**:262-273. DOI: 10.1016/j.fbp.2020.07.006

[63] Ambrosi A, Motke MB, Souza-Silva ÉA, Zini CA, McCutcheon JR, Cardozo NSM, et al. Beer dealcoholization by forward osmosis diafiltration. Innovative Food Science and Emerging Technologies. 2020;**63**: 102371. DOI: 10.1016/j.ifset.2020.102371

[64] Massot A, Mietton-Peuchot M, Peuchot C, Milisic V. Nanofiltration and reverse osmosis in winemaking. Desalination. 2008;**231**:283-289. DOI: 10.1016/j.desal.2007.10.032

[65] Sam FE, Ma T, Liang Y, Qiang W, Atuna RA, Amagloh FK, et al. Comparison between membrane and thermal dealcoholization methods: Their impact on the chemical parameters,

volatile composition , and sensory characteristics of wines. Membranes (Basel). 2021;**11**:957

[66] Bui K, Dick R, Moulin G, Galzy P. Partial concentration of red wine by reverse osmosis. Journal of Food Science. 1988;**53**:647-648. DOI: 10.1111/j.1365-2621.1988.tb07776.x

[67] López M, Alvarez S, Riera FA, Alvarez R. Production of low alcohol content apple cider by reverse osmosis. Industrial and Engineering Chemistry Research. 2002;**41**:6600-6606. DOI: 10.1021/ie020155a

[68] Varavuth S, Jiraratananon R, Atchariyawut S. Experimental study on dealcoholization of wine by osmotic distillation process. Separation and Purification Technology. 2009;**66**: 313-321. DOI: 10.1016/j.seppur.2008.12.011

[69] Diban N, Arruti A, Barceló A, Puxeu M, Urtiaga A, Ortiz I. Membrane dealcoholization of different wine varieties reducing aroma losses. Modeling and experimental validation. Innovative Food Science and Emerging Technologies. 2013;**20**:259-268. DOI: 10.1016/j.ifset.2013.05.011

[70] Hogan P, Canning R, Peterson P, Johnson R, Michaels A. A new option : Osmotic distillation. Chemical Engineering Progress. 1998;**94**:49-61

[71] Liguori L, Attanasio G, Albanese D, Di Matteo M. Aglianico Wine Dealcoholization Tests. In Computer Aided Chemical Engineering. Amsterdam, The Netherlands: Elsevier. 2010;**28**:325-330

[72] Belisario-Sánchez YY, Taboada-Rodríguez A, Marín-Iniesta F, Iguaz-Gainza A, López-Gómez A. Aroma Recovery in Wine Dealcoholization by SCC Distillation. Food and Bioprocess Technology. 2012;**5**:2529-2539. DOI: 10.1007/s11947-011-0574-y

[73] Liguori L, Russo P, Albanese D, Di Matteo M. Evolution of quality parameters during red wine dealcoholization by osmotic distillation. Food Chemistry. 2013;**140**:68-75. DOI: 10.1016/j.foodchem.2013.02.059

[74] Motta S, Guaita M, Petrozziello M, Ciambotti A, Panero L, Solomita M, et al. Comparison of the physicochemical and volatile composition of wine fractions obtained by two different dealcoholization techniques. Food Chemistry. 2017;**221**:1-10. DOI: 10.1016/j.foodchem.2016.10.046

[75] Margallo M, Aldaco R, Barceló A, Diban N, Ortiz I, Irabien A. Life cycle assessment of technologies for partial dealcoholisation of wines. Sustainable Production and Consumption. 2015;**2**: 29-39. DOI: 10.1016/j.spc.2015.07.007

[76] Catarino M, Mendes A. Non-alcoholic beer—A new industrial process. Separation and Purification Technology. 2011;**79**:342-351. DOI: 10.1016/j.seppur.2011.03.020

[77] Pham DT, Ristic R, Stockdale VJ, Jeffery DW, Tuke J, Wilkinson K. Influence of partial dealcoholization on the composition and sensory properties of Cabernet Sauvignon wines. Food Chemistry. 2020;**325**:126869. DOI: 10.1016/j.foodchem.2020.126869

[78] Salgado CM, Fernández-Fernández E, Palacio L, Carmona FJ, Hernández A, Prádanos P. Application of pervaporation and nanofiltration membrane processes for the elaboration of full flavored low alcohol white wines. Food and Bioproducts Processing. 2017;**101**:11-21. DOI: 10.1016/j.fbp.2016.10.001

[79] Pham DT, Stockdale VJ, Wollan D, Jeffery DW, Wilkinson KL. Compositional consequences of partial dealcoholization of red wine by reverse osmosis-evaporative perstraction. Molecules. 2019;**24**:1-16. DOI: 10.3390/molecules24071404

[80] Russo P, Liguori L, Corona O, Albanese D, Matteo MD, Cinquanta L. Combined membrane process for dealcoholization of wines: Osmotic distillation and reverse osmosis. Chemical Engineering Transactions. 2019;**75**:7-12. DOI: 10.3303/CET1975002

[81] Ma TT, Sun XY, Gao GT, Wang XY, Liu XY, Du GR, et al. Phenolic characterisation and antioxidant capacity of young wines made from different grape varieties grown in helanshan donglu wine zone (China). South African Journal of Enology and Viticulture. 2014;**35**:321-331. DOI: 10.21548/35-2-1020

[82] Suo H, Tian R, Li J, Zhang S, Cui Y, Li L, et al. Compositional characterization study on high - molecular -mass polymeric polyphenols in red wines by chemical degradation. Food Research International. 2019;**123**: 440-449. DOI: 10.1016/j.foodres.2019.04.056

[83] Delgado Cuzmar P, Salgado E, Ribalta-Pizarro C, Olaeta JA, López E, Pastenes C, et al. Phenolic composition and sensory characteristics of Cabernet Sauvignon wines: effect of water stress and harvest date. International Journal of Food Science and Technology. 2018;**53**: 1726-1735. DOI: 10.1111/ijfs.13757

[84] Gómez-Plaza E, López-Nicolás JM, López-Roca JM, Martínez-Cutillas A. Dealcoholization of Wine. Behaviour of the Aroma Components during the Process. LWT—Food Science and Technology. 1999;**32**:384-386. DOI: 10.1006/fstl.1999.0565

[85] Frost R, Quiñones I, Veldhuizen M, Alava JI, Small D, Carreiras M. What can the brain teach us about winemaking? An fMRI study of alcohol level preferences. PLoS One. 2015;**10**:1-11. DOI: 10.1371/journal.pone.0119220

[86] Escudero A, Campo E, Fariña L, Cacho J, Ferreira V. Analytical characterization of the aroma of five premium red wines. Insights into the role of odor families and the concept of fruitiness of wines. Journal of Agricultural and Food Chemistry. 2007;**55**:4501-4510. DOI: 10.1021/jf0636418

[87] Bellsario-Sánchez YY, Taboada-Rodríguez A, Marín-Iniesta F, López-Gómez A. Dealcoholized wines by spinning cone column distillation: Phenolic compounds and antioxidant activity measured by the 1,1-Diphenyl-2-picrylhydrazyl method. Journal of Agricultural and Food Chemistry. 2009; **57**:6770-6778. DOI: 10.1021/jf900387g

[88] Lisanti MT, Gambuti A, Genovese A, Piombino P, Moio L. Partial Dealcoholization of Red Wines by Membrane Contactor Technique: Effect on Sensory Characteristics and Volatile Composition. Food and Bioprocess Technology. 2013;**6**:2289-2305. DOI: 10.1007/s11947-012-0942-2

[89] Flanzy C. Enología: fundamentos científicos y tecnológicos; AMV Edicione. Madrid: Mundi-Prensa; 2003

[90] Asproudi A, Petrozziello M, Cavalletto S, Guidoni S. Grape aroma precursors in cv. Nebbiolo as affected by vine microclimate. Food Chemistry. 2016;**211**:947-956. DOI: 10.1016/j.foodchem.2016.05.070

[91] Meillon S, Urbano C, Schlich P. Contribution of the temporal dominance of sensations (TDS) method to the sensory description of subtle differences in partially dealcoholized red wines. Food Quality and Preference. 2009;**20**:490-499. DOI: 10.1016/j.foodqual.2009.04.006

[92] King ES, Heymann H. The effect of reduced alcohol on the sensory profiles and consumer preferences of white wine. Journal of Sensory Studies. 2014; **29**:33-42. DOI: 10.1111/joss.12079

[93] Meillon S, Viala D, Medel M, Urbano C, Guillot G, Schlich P. Impact of partial alcohol reduction in Syrah wine on perceived complexity and temporality of sensations and link with preference. Food Quality and Preference. 2010;**21**:732-740. DOI: 10.1016/j.foodqual.2010.06.005

[94] Sánchez-Palomo E, Delgado JA, Ferrer MA, Viñas MAG. The aroma of La Mancha Chelva wines: Chemical and sensory characterization. Food Research International. 2019;**119**:135-142. DOI: 10.1016/j.foodres.2019.01.049

[95] Duan WP, Zhu BQ, Song RR, Zhang B, Lan YB, Zhu X, et al. Volatile composition and aromatic attributes of wine made with vitisvinifera l.Cv cabernet sauvignon grapes in the xinjiang region of china: Effect of different commercial yeasts. International Journal of Food Properties. 2018;**21**:1423-1441. DOI: 10.1080/10942912.2018.1479860

[96] Diban N, Athes V, Bes M, Souchon I. Ethanol and aroma compounds transfer study for partial dealcoholization of wine using membrane contactor. Journal of Membrane Science. 2008;**311**:136-146. DOI: 10.1016/j.memsci.2007.12.004

[97] Tonutti I, Liddle P. Aromatic plants in alcoholic beverages. A review. Flavour and Fragrance Journal. 2010;**25**:341-350. DOI: 10.1002/ffj.2001

[98] Styger G, Prior B, Bauer FF. Wine flavor and aroma. Journal of Industrial Microbiology & Biotechnology. 2011;**38**: 1145-1159. DOI: 10.1007/s10295-011-1018-4

[99] Jordão AM, Vilela A, Cosme F. From sugar of grape to alcohol of wine: Sensorial impact of alcohol in wine. Beverages. 2015;**1**:292-310. DOI: 10.3390/beverages1040292

[100] Catarino M, Mendes A. Dealcoholizing wine by membrane separation processes. Innovative Food Science and Emerging Technologies. 2011;**12**:330-337. DOI: 10.1016/j.ifset.2011.03.006

[101] Liang Z, Zhang P, Zeng XA, Fang Z. The art of flavored wine: Tradition and future. Trends in Food Science and Technology. 2021;**116**:130-145. DOI: 10.1016/j.tifs.2021.07.020

[102] Parker M, Barker A, Black CA, Hixson J, Williamson P, Francis IL. Don't miss the marc: phenolic-free glycosides from white grape marc increase flavour of wine. Australian Journal of Grape and Wine Research. 2019;**25**:212-223. DOI: 10.1111/ajgw.12390

[103] Pedroza MA, Carmona M, Alonso GL, Salinas MR, Zalacain A. Pre-bottling use of dehydrated waste grape skins to improve colour, phenolic and aroma composition of red wines. Food Chemistry. 2013;**136**:224-236. DOI: 10.1016/j.foodchem.2012.07.110

[104] Culea RE, Tamba-berehoiu RM, Popa NC. Sensory properties of some white wines, flavored wines and vermouth type wines, prepared by using own recipes. Scientific Papers Series Management,